SpringerBriefs in Environmental Science

SpringerBriefs in Environmental Science present concise summaries of cutting-edge research and practical applications across a wide spectrum of environmental fields, with fast turnaround time to publication. Featuring compact volumes of 50 to 125 pages, the series covers a range of content from professional to academic. Monographs of new material are considered for the SpringerBriefs in Environmental Science series.

Typical topics might include: a timely report of state-of-the-art analytical techniques, a bridge between new research results, as published in journal articles and a contextual literature review, a snapshot of a hot or emerging topic, an in-depth case study or technical example, a presentation of core concepts that students must understand in order to make independent contributions, best practices or protocols to be followed, a series of short case studies/debates highlighting a specific angle.

SpringerBriefs in Environmental Science allow authors to present their ideas and readers to absorb them with minimal time investment. Both solicited and unsolicited manuscripts are considered for publication.

More information about this series at http://www.springer.com/series/8868

Richard Meissner

Interest Groups, Water Politics and Governance

The Case of the Lesotho Highlands Water Project

Springer

Richard Meissner
Centre for Water Resources Research
University of KwaZulu-Natal
Scottsville
South Africa

ISSN 2191-5547 ISSN 2191-5555 (electronic)
SpringerBriefs in Environmental Science
ISBN 978-3-319-21129-9 ISBN 978-3-319-21130-5 (eBook)
DOI 10.1007/978-3-319-21130-5

Library of Congress Control Number: 2015942788

Springer Cham Heidelberg New York Dordrecht London

Printed on acid-free paper

Springer International Publishing AG Switzerland is part of Springer Science+Business Media (www.springer.com)

Acknowledgments

This book is the result of research conducted for my D.Phil. dissertation entitled *The Transnational Role and Involvement of Interest Groups in Water Politics: A Comparative Analysis of Selected Southern African Case Studies*, which I completed in 2004 at the University of Pretoria. I would like to thank Prof. Anton du Plessis from the Department of Political Sciences at the University of Pretoria for his supervision of the thesis. Without his assistance, the thesis and this book would not have been possible. I would also like to thank Prof. Maxi Schoeman for kindly agreeing to the production of this book from my thesis. Dr. Marius Claassen my Research Group Leader, at the time, at the Council for Scientific and Industrial Research also deserves a word of thanks for his encouragement and support. Since I conducted the research nearly a decade ago, I had to update the research, and for this, I would like to thank a number of individuals.

Contents

Chapter 1
Introduction

Abstract Interest groups are pervasive phenomena of modern society. Although this is so, water governance researchers generally do not include them in their research endeavours. This is especially the case regarding research on transboundary rivers. Interest groups have been an opposing force to large dams the world over. The southern African region is no exception. In the region, and South Africa in particular, water is made available through water engineering projects. One of which is the Lesotho Highlands Water Project (LHWP). In this book, I investigate the transnational role and involvement of interest groups in the LHWP.

Keywords Interest groups · Transnational politics · Lesotho Highlands Water Project · Water discourse · Transboundary river · Orange-Senqu River

Interest groups are pervasive phenomena of modern society. The attitude of democratic governments towards interest groups is reactive in nature in that they respond to the desires of the people, and interest groups are important actors in this relationship. More specifically, they are one of the instruments through which people can express their political desires. In short, interest groups act as conduits for citizens to communicate with government (Rothenberg 1992; Sadie 1998; Hunter 1999; Strolovitch 2006) and to realise objectives governments are unable to provide. In most issues in domestic or international politics, interest groups have come to represent the desires of people internal and external to their constituencies. These issues range from health care to the provision of wholesome freshwater. Not only do interest groups represent people's desires, but they also play an important role in mobilising support to either change or enhance policies governing issues. This also applies to water resource management projects such as large dams. Over the last three and a half centuries, the state has played an important role in the implementation of such projects through the so-called hydraulic mission (Reisner 1993). This is not the case anymore, with a global anti-dam movement engaging governments and inter-governmental organisations (IGOs) over the issue of large dams' impacts on humans and the environment.

R. Meissner, *Interest Groups, Water Politics and Governance*,
SpringerBriefs in Environmental Science, DOI 10.1007/978-3-319-21130-5_1

The anti-dam debate has been raging for some years in developed countries, particularly the USA and Europe. This debate has spilled over into the developing world and has nestled among sections of the public that are opposed to such plans (McCully 1996; Rothman and Oliver 1999; Biswas 2004; McCormick 2006; Peterson 2010). Southern Africa's transboundary rivers are no exception. Internal and external to the region, a transnational movement against projects has been in existence for quite some time. Regardless of this, and in contrast to the USA and Europe, Latin America, and India, there is a paucity of international relations research on southern African cases.

That being the case, interest groups are becoming an ever greater, opposing dynamism to contend with when governments construct water projects on southern African international rivers. Examples to illustrate this are the Lesotho Highlands Water Project across the upper reaches of the Orange-Senqu River.

In South Africa, freshwater is made possible through water projects, such as large dams and elaborate irrigation schemes supplying bulk water to utilities which in turn provide it to local governments and then to the public (Turton and Meissner 2002). Stated differently, supplying water throughout this process is accomplished with policies to ensure that the largest number of people has adequate sanitation and potable water. Through these policies, states have become the custodians of the water resources in, or flowing across, their territories. Custodianship is traditionally imbedded in the principles of sovereignty and territorial integrity. Because the state is responsible for the well-being of the population, interest groups lobbying against water projects come into direct opposition with government policy to implement these projects. This implies that as interest groups lobby against these projects, there might be an 'erosion' of the state's authority to construct water projects.

Interest groups perform a specific function in society concerning such projects. They advocate the cause of people and the environment affected by major water projects and oppose these schemes. Moreover, these non-state entities consider it their responsibility to stop water development projects and they will do anything in their power to do so. Phrased in another way, some interest groups believe that water projects have severe implications for aquatic *fauna* and *flora*, and that some plant and animal species have even become rare or extinct because of these projects. Therefore, a social and environmental conscience or awareness has developed in recent decades to protect both the environment and the humans from the adverse effects of these schemes, and interest groups are at the forefront of this normative endeavour (Gleick 1998) both domestically and across state boundaries.

As water becomes scarcer across the world, many states in the developing world are looking towards water resources management projects to solve their water deficiency problems and energy needs. We should remember that such schemes do not only provide bulk water, but can also generate much needed hydroelectricity. As articulated and expressed by interest groups, there is an increasing awareness of the negative impacts of water projects on communities and the environment. Since the early 1990s, engineers, water managers, and decision-makers in the water sector have been facing increased opposition towards such plans. The opposition from interest groups is generally restricting the policy choices of governments, project planners, and implementers of such schemes. This has already led to a ballooning political interaction between states and interest groups and in all likelihood will continue to increase in the future around the issue of water development projects.

Researchers, consultants, or academics in the water discourse do not sufficiently deal with the (transnational) role and involvement of interest groups. A number of researchers have discussed the transnational role and involvement of interest groups in water politics, to a certain extent and either explicitly or implicitly. They are Payne (1996[1]), McCully (1996[2]), Gleick (1998[3]), Meissner (1998a, b[4], 2000a, b, c, d, 2003), Turton and Meissner (2002), and Kgomongo and Meissner (2004). Apart from this, some of these research endeavours adopt a state-centric approach,[5] for example Teclaff (1967), Naff and Matson (1984), Lowi (1993), Gleick (1993), Kliot (1994), Hillel (1994), Meissner (1998b), Ashton (2000), Turton (2000), and Jägerskog (2002). This has implications for the theoretical relevance of the study. Water governance researchers generally do not study interest groups as one of the main actors within the water discourse but relegate these actors to second position in their research endeavours. What is furthermore not considered is how, where, and when they might reduce the state's ability to construct water projects. This implies that interest groups are rather studied as actors that are part of a broad array of entities involved in water politics, and not as separate agents. In other words, they are not always the central focus of practical research in the field of water politics.

Research on interest group involvement in any aspect of the river basin opens the possibility that an interpretivist approach as opposed to a positivist slant (e.g. Conley 1995, 1996; Meissner 1998b, 2004, 2005a, b, c, 2010; Jägerskog 2003; Heyns 2003; Meissner et al. 2012; Mwenge Kahinda et al. 2012; Turton 2005; Earle

[1]Payne's (1996) article examines the campaigning of interest groups against large dams in that they do not only target states but other international governmental bodies in the international political realm, e.g. the World Bank.

[2]McCully (1996) looks at the politics and ecological implications large dams have on the environment and people, with the last chapter of his book devoted to the role of interest groups in various countries in the world lobbying against these structures.

[3]In his book, Gleick examines the role and activities of interest groups with respect to their lobbying against large dams within a global context.

[4]The subsequent articles of Meissner analyses a number of cases where interest groups campaigned against WRMPs within the Southern African context, like Namibia's Okavango Water Pipeline.

[5]These studies look at various aspects of water politics especially where the state has played a prominent role. For instance, Teclaff's book studied the international river basin from a historical and international law perspective, where the state has been at least one of the parties within bilateral and multilateral agreements concerning the use of national and international rivers. The other studies looked at international rivers, in various regions such as the Middle East, Southern Africa and Southeast Asia, from a conflict and cooperative perspective. In other words, these studies had, as their central focus, the state as the main actor concerning the water politics within the different regions under consideration.

et al. 2005; Davidsen 2006; DWAF 2009; and Jacobs 2009, 2012) is followed. In the following sections of this book, I attempt to do just that: investigate the role and involvement of interest groups in the LHWP from an interpretivist perspective.

References

Ashton, P. 2000. *Southern Africa water conflicts: Are they inevitable or are they preventable?* Geneva: Green Cross International.

Biswas, A.K. 2004. Dams: cornucopia or disaster? *International Journal of Water Resources Development* 20(1): 3–14.

Conley, A.H. 1995. A synoptic view of water resources in Southern Africa. Paper presented at the conference of Southern Africa foundation for economic research on integrated development of regional water resources, in Nyanga, Zimbabwe.

Conley, A.H. 1996. The need to develop the water resources of Southern Africa. Paper presented at the Vic Falls conference on aquatic systems, July 1996, in Zimbabwe.

Davidsen, P.A. 2006. *The making and unmaking of the politics of exceptionality: Studying processes of securitisation and desecuritisation in the Orange and Okavango River basins.* Thesis submitted for the degree Master of Comparative Politics, Institute of Comparative Politics, The University of Bergen.

Department of Water Affairs and Forestry (DWAF). 2009. *Orange River: Assessment of water quality data requirements for planning purposes. Resource Water Quality Objectives (RWQOs): Upper and Lower Orange Water Management Areas (WMAs 13 and 14).* Report no 5 (P RSA D000/00/8009/2). Pretoria: Department of Water Affairs and Forestry.

Earle, A., D. Malzbender, A. Turton, and E. Manzungu. 2005. *A preliminary basin profile of the Orange/Senqu River*. Pretoria: African Water Issues Research Unit at the University of Pretoria.

Gleick, P.H. 1993. Water and conflict: fresh water resources and international security. *International Security* 18(1): 84–117.

Gleick, P.H. 1998. *The world's water 1998–1999: The biennial report on freshwater resources*. Washington, D.C.: Island Press.

Heyns, P. 2003. Water-resources management in Southern Africa. In *International waters in Southern Africa*, ed. M. Nakayama. Tokyo: United Nations University Press.

Hillel, D. 1994. *Rivers of Eden: The struggle for water and the quest for peace in the Middle East*. Oxford: Oxford University Press.

Hunter, K.G. 1999. *Interest group and state economic development*. Westport, CT: Praeger Publishers.

Jacobs, I.M. 2009. *Norms and transboundary co-operation in Africa: The cases of the Orange-Senqu and Nile Rivers*. PhD Dissertation: University of St. Andrews, St. Andrews, Scotland.

Jacobs, I.M. 2012. A community in the Orange: the development of a multi-level water governance framework in the Orange-Senqu River basin in Southern Africa. *International Environmental Agreements: Politics, Law and Economics* 12: 187–210.

Jägerskog, A. 2002. Contributions of regime theory in understanding interstate water co-operation: Lessons learned in the Jordan River basin. In *Hydropolitics in the developing world: A Southern African perspective*, ed. A. Turton, and R. Henwood. Pretoria: African Water Issues Research Unit (AWIRU).

Jägerskog, A. 2003. *Why states cooperate over shared water: The water negotiations in the Jordan River basin*. Linköping, Sweden: Department of Water and Environmental Studies, Linköping University.

Kgomongoe, M., and R. Meissner. 2004. The revised protocol on shared watercourses and the management of water resources in SADC. In *Yearbook of international affairs 2003/04*, ed. South African. Johannesburg: South African Institute of International Affairs.

Kliot, N. 1994. *Water resources and conflict in the Middle East*. London and New York: Routledge.

Lowi, M.R. 1993. *Water and power: The politics of a scarce resource in the Jordan River basin*. Cambridge: Cambridge University Press.

McCormick, S. 2006. The Brazilian anti-dam movement: Knowledge contestation as communicative action. *Organization and Environment* 19: 321–346.

McCully, P. 1996. *Silenced rivers: The ecology and politics of large dams*. London: Zed Books.

Meissner, R. 1998a. Piping water from the Okavango River to Namibia—The role of communities and pressure groups in water politics. *Global Dialogue*, 3(2).

Meissner, R. 1998b. *Water as 'n bron van politieke konflik en samewerking: 'n Vergelykende studie van die Midde-Ooste and Suider-Afrika*. Tesis vir die graad Magister Artium in Lettere en Wysbegeerte, Randse Afrikaanse Universiteit, Johannesburg.

Meissner, R. 2000a. The case of the Orange River. *Conflict Trends* 2: 24–27.

Meissner, R. 2000b. In the spotlight…Interest groups as a role player in large water projects. *SA Waterbulletin* 26(2): 24–26.

Meissner, R. 2000c. *And now time for some good news: Cooperation in the Orange River basin*. Pretoria: African Water Issues Research Unit.

Meissner, R. 2000d. Hydropolitical hotspots in Southern Africa: The case of the Kunene River. In *Water wars: Enduring myth or impending reality*, ed. H. Solomon, and A. Turton. Durban: The African Centre for the Constructive Resolution of Disputes (ACCORD).

Meissner, R. 2003. Interaction and existing constraints in international river basins. In *International waters in Southern Africa*, ed. M. Nakayama. Tokyo: United Nations University Press.

Meissner, R. 2004. *The transnational role and involvement of interest groups in water politics: A comparative analysis of selected Southern Africa case studies*. D.Phil dissertation, University of Pretoria, Faculty of Humanities.

Meissner, R. 2005a. Die politiek van die ekologie: Earthlife Africa en die President: Belangegroepe as rolspelers in die samelewing. *Die Vrye Afrikaan*, 2(16).

Meissner, R. 2005b. Interest groups and the proposed Epupa Dam: Towards a theory of water politics. *Politeia* 24(3): 354–370.

Meissner, R. 2005c. Interest groups as local stakeholders involved in the water politics of a transboundary river: The case of the proposed Epupa Dam across the Kunene River. In Water, development and cooperation—Comparative perspective: Euphrates-Tigris and Southern Africa, ed. L. Wirkus, Paper 46. Bonn: Bonn International Center for Conversion.

Meissner, R. 2010. The water security and business nexus on the African continent. *Africa Industry & Business*, 16 June 2010.

Meissner, R., Claassen, M., Funke, N., Ntombela, C. 2012. Proceedings of workshops focusing on various aspects of South African water politics, governance and institutional arrangements. Pretoria: Council for Scientific and Industrial Research. Publication number: CSIR/NRE/ECOS/IR/2012/0010/C.

Mwenge Kahinda, J., Kapangaziwiri, E., Engelbrecht, F., Meissner, R., Ashton, P.J. 2012. *Implementing integrated catchment management in the Limpopo River Basin. Phase 1: Situational assessment*. Pretoria: Council for Scientific and Industrial Research.

Naff, T., and R.C. Matson. 1984. *Water in the Middle East: Conflict or cooperation*. Boulder, Colorado: Westview Press.

Payne, R.A. 1996. Non-profit environmental organizations in world politics: Domestic structure and transnational relations. *Policy Studies Review*, 14 (Spring/Summer 1995): 171–182.

Peterson, M.J. 2010. *Narmada dams controversy—Case summary*. International Dimensions of Ethics Education in Science and Engineering. Amherst: University of Massachusetts.

Reisner, M. 1993. *Cadillac desert: The American West and its disappearing water*. New York: Penguin Books.

Rothenberg, L.S. 1992. *Linking citizens to government: Interest group politics at common cause*. Cambridge: Cambridge University Press.
Rothman, F.D., and P.E. Oliver. 1999. From local to global: The anti-dam movement in southern Brazil, 1979–1992. *Mobilization: An International Journal* 4(1): 41–57.
Sadie, Y. 1998. Political parties and interest groups. In *Government and politics in the new South Africa*, ed. A.J. Venter. Pretoria: J.L. van Schaik.
Strolovitch, D.Z. 2006. Do interest groups represent the disadvantaged? Advocacy at the intersection of race, class, and gender. *The Journal of Politics* 68(4): 894–910.
Teclaff, L.A. 1967. *River basin in history and law*. Amsterdam: Martinus Nijhoff.
Turton, A.R. 2000. Precipitation, people, pipelines and power in Southern Africa. In *Political ecology: Science, myth and power*, ed. P. Stott, and S. Sullivan. London: Arnold.
Turton, A. 2005. *Hydro hegemony in the context of the Orange River Basin*. Paper presented at the Workshop on Hydro Hegemony, Hosted by Kings College and the School of Oriental and African Studies, May 20–21, 2005, in London.
Turton, A., and R. Meissner. 2002. The hydrosocial contract and its manifestation in society: A South African case study. In *Hydropolitics in the developing world: A Southern African perspective*, ed. A. Turton, and R. Henwood. Pretoria: The African Water Issues Research Unit (AWIRU).

Chapter 2
Interest Groups

Abstract Interest groups are not like units; they have different characteristics and do not only involve themselves in domestic politics, but also international politics. There are various definitions of interest groups. I argue that there is a difference between a non-governmental organisation and an interest group. A non-governmental organisation performs a service, whereas an interest group influences public and private policies. When interest groups from one country influence the policies of another country's government or private institutions, they are acting in a transnational manner. In this chapter, I move away from the predominantly positivist paradigm and argue for an interpretivist stance towards researching interest groups. To do so, I employ analytic eclecticism as a point of departure. Doing research from an interpretivist stance promises a more nuanced understanding of the transnational role and involvement of interest groups in water engineering projects.

Keywords Conceptualisation · Interpretivism · Analytic eclecticism · Everyday international political economy · Paradigm

Introduction

The purpose of this section is, firstly, to outline the fundamental characteristics of interest groups and, secondly, to discuss the nature and extent of interest groups' transnational activities. Of importance in this chapter is not to debate which of the different conceptualisations is correct, but to investigate how interest groups stand in relation to states, their governing bodies and other non-state entities in society. To accomplish this, the chapter will define the concept interest group and related entities such as non-governmental organisations (NGOs) and pressure groups; provide a typology of interest groups; discuss these political entities as transnational actors; highlight their authority roles; briefly discuss a different perspective at investigating interest groups; and lastly draw a conclusion.

R. Meissner, *Interest Groups, Water Politics and Governance*,
SpringerBriefs in Environmental Science, DOI 10.1007/978-3-319-21130-5_2

Defining Interest Groups

Various definitions of the concept interest group exist. Within the Political Sciences, researchers use these concepts in a confusing manner. For instance, not all scholars apply the same meaning to the concept *interest group*; they define it differently.

As a point of departure, throughout the twentieth century, studies on interest group politics were frequently conducted, and consequently, different terms to describe interest groups developed. Nevertheless, a generic definition would state that an interest group is a non-state entity that endeavours to influence government policy. In addition, in this book, the notion of 'interest' will be used to connote the pursuit of causes or the advancement and defence of particular interests, positions, and people in society (Grant 2000). Even so, phrases, other than *interest group*, include *pressure group*, *lobby*, *NGO*, and the variant of the latter *international non-governmental organisation (INGO)*. How have scholars defined these entities? In this book, I take the stance that *interest groups*, *pressure groups*, *NGOs/INGOs*, and *lobbies* are not identical. All interest groups are per definition NGOs, but not all NGOs are interest groups. Non-governmental organisations only present the interests of their members, but may not act to influence government policies. It is only when an NGO starts to influence governmental policies that it becomes an interest group. For instance, the Automobile Association (AA) is an NGO that provides a service to its members. That being said, when it starts to make statements on how the death toll on South African roads can be lowered, through 'better' governmental policies, it becomes an interest group. Similarly, all pressure groups and lobbies are interest groups, but not all interest groups are pressure groups and lobbies. Pressure groups and lobbies focus their activities mainly on governmental policies or the governmental process. Interest groups, on the other hand, attempt to influence government policies, the governmental process, *and* other organisations in society through various means. For instance, the Treatment Action Campaign (TAC) is an interest group, influencing the South African government to change its policies on HIV/AIDS-infected persons. At the same time, it also influences pharmaceutical companies to lower their prices on antiretroviral drugs.

All these concepts, except *NGO/INGO*, have been treated as something of a profanity in the scholarly discourse. Many scholars, members of interest groups, and government officials prefer one concept to the other because of the negative connotations that can be attached to some of the concepts. As far as Truman (1951: 38) is concerned, the concept pressure group invokes a sense of 'selfishness' and that 'special rights and privileges' are sought by these groups from other political actors. Similarly, Moulder et al. (1982) prefer to use interest group because pressure group implies negativity towards these organisations. One of the reasons for this negative image of pressure groups, as Mackenzie (1955) declares, is that pressure always invokes power politics. In contrast, diplomats and government officials prefer to use the concept 'NGO'. The reason for this is that ambassadorial representatives and politicians like to allege that they aspire to seek and represent the national interest of a united society. Furthermore, they will not concede that they are standing in a relationship with interest groups (Willets 1997). Even so, over the years, the concept 'interest group' has been used with wantonness, which caused unnecessary disorder about its actual meaning (Geldenhuys 1998).

It follows then that a common denominator that is employed in almost all definitions is 'influence on governmental/public policies'. Traditionally, influence is a component of power (Holsti 1995: 118) and is related to the actions of interest groups within their relationship with government or any other entity like an international organisation or financial institution.

Various scholars have defined interest groups around the notion of influence either very narrowly or to a wider extent. Petracca (1992) observes that, currently, academics are more likely to use concepts that are pertinent to their research endeavours. Baumgartner and Leech (1998: xxii) insist that: 'When we refer to "interest groups" or "organised interests", we mean not only membership organizations that do not accept members, businesses, and any other organisation or institution that makes policy-related appeals to government'. Baumgartner and Leech (1998) do not have a 'strong preference' towards one of a number of phrases. What is of more importance, than to argue over the different concepts, is to point to the role interest groups play in society and more appropriately their relationship with the state and government officials and other non-state entities.

For Baumgartner and Leech (1998: xxii), a concept makes no difference, for 'we are pluralists on this score'. This sentiment is echoed in other studies as well. Smith (1993: 1) states that, '[t]he important variables in understanding decisions are the nature of the relationships that exist between groups and the state—the types of policy networks—and the interest and activities of state actors—the degree of autonomy'. Baumgartner and Leech (1998) and Smith (1993) therefore note that, rather than broaden the debate over concepts, the relationship that exists between interest groups, the state and government, their impact on state autonomy, and policy issues should be the focus.

In recent years, political scientists have moved away from the narrow thinking of interest groups, in the sense that they represent the interests of a specific group in society. Increasingly, interest groups have been defined with their actions towards the state and other organisations, with influence or advocacy playing a central role. The study of interest groups may bring scholars closer to answering questions concerning the relationship between the state and society (Wilson 1990). In this book, I will employ the concept interest group although there is a tendency in International Relations literature to use the phrase *NGO* more frequently. Taken a step further, some scholars do not grapple with the difficulty of the terminology. As has been indicated, they select a concept and look at the real issue in group politics, namely the role and purpose of interest groups, the processes involved, and the relationship of interest groups with the state and other governmental and non-governmental actors. With this in mind: how do scholars in the water discourse threat the concept interest group?

Within the water discourse, a number of scholars prefer to use the concept *NGO* or *stakeholder* when writing or referring to these organisations, for instance McCully (1996), Gleick (1998), Turton and Ohlsson (1999), and Turton (2000b). Is this wrong? Most certainly not. The scholars and practitioners in the water discourse refer to a broad range of organisations and not only to specific institutions when they write about non-state actors in water politics. It should also be kept in mind that not all these scholars are Political Scientists or International Relations specialists. Some analysts within Political Sciences are, like their non-Political Science colleges, unfamiliar with the concepts such as *interest group*, *pressure group,* and *lobby*. They therefore do not present clear definitions within the rubric of their research. In addition, at times they do not focus on interest groups in water politics. Some are experts within the field of water politics and may not have come across these concepts in their work. Many practitioners and decision-makers are also not accustomed to the concepts used in the analysis of group politics. An example of this is the former director general of South Africa's Department of Water Affairs and Forestry (DWAF) using the concept *NGO* in referring to entities lobbying against large dams (M. Muller, personal communication, 13 February 2001). Throughout the book when I talk about an interest group, I mean that it is a non-state entity that influences government policies and other non-state and intergovernmental institutions, in the national political and the international political domain.

Interest Groups as Transnational Actors

International Relations as a field of study is often understood to encompass the associations between states. Other actors in world politics such as economic institutions [that is multinational corporations (MNCs) and social groups] are in many cases given a secondary status as non-state entities in world affairs (Willets 1997). This is in line with the realist perspective of world politics, but not with liberal pluralism. This latter perspective suggests that non-state actors, like interest groups, also play roles in the international system, although they are confined to a second tier of analysis. Even so, interest groups play an important role within world politics when operating as transnational actors and should not be confined to a second tier of analysis.

In this regard, Archer (1992: 1) defines transnational relations as the 'activities between individuals and groups in one state and individuals and groups in another state'. Another definition is that it is the 'contacts, coalitions, and interactions across State boundaries that are not controlled by the central foreign policy organs of government' (Nye and Keohane 1971: 330).

Since interest groups do not confine their activities to one state or to domestic policy issues only, they play a transnational role by lobbying governments across state borders. An interest group from a particular state is able to cross political boundaries and start a lobbying campaign within another country. In this sense, interest groups can intervene in policy arenas at a transnational level, to promote the interests of their members, provide assistance, and promote scientific and technological research and the communication and advancement of ideological, cultural, and religious ideas (Ghils 1992). Stated differently, interest groups, like states, can cross international boundaries, communicate with each other, and set up networks or even alliances through which efforts are coordinated. In effect, interest groups, as transnational actors, can be considered older than the modern state.

In this regard, before the formation of the Westphalian state, in 1648, transnational actors, like trading organisations such as the British East India Company, played a leading role in the spread of ideas and ideologies. The state was not the only independent actor in world politics. Because of the spreading of ideas and ideologies, social, political, economic, and cultural actors 'had to be subnational, transnational or supranational' (Stern 2000: 246). Regarding ideas, non-state entities always had a transnational character in terms of the spread and formation of non-state actors. Stated more precisely, whenever political actors interacted, 'assumptions, concepts, creeds, doctrines and dogmas could always be transmitted from one to another' (Stern 2000: 246).

This is still the case in the contemporary era with relations between non-state entities driven by the process of globalisation.[1] Globalisation 'refers to processes whereby social relations acquire relatively distance-less and borderless qualities, so that human lives are increasingly played out in the world as a single place. Social relations—that is, the countless and complex ways that people interact with and affect each other—are more and more being conducted and organized on the basis of a planetary unit' (Scholte 1999: 14–15). Globalisation therefore has the effect that social relations are to an increasing extent conducted at a broader level than at the time before the Westphalian state. Thus, and because of globalisation, the world has become a relatively borderless social environment in which interest groups are operating (Scholte 1999).

Interest groups have also stake a claim in their dealings with the so-called main actors in world affairs—states. In this regard, Heywood (1997: 265) asserts that 'in the closing decades of the twentieth century, interest group activity has increasingly adjusted to the impact of globalisation and the strengthening of supranational bodies. Amongst the groups best suited to take advantage of such shifts are charities and environmental campaigners (such as Greenpeace and Friends of the Earth) which already possess transnational structures and an international membership'. One of the possible reasons why interest groups, with an environmental agenda, are so adept to conform to globalisation trends is the nature of environmental problems facing humankind. Environmental problems are not localised to a single country. Many environmental problems cross state boundaries affecting regions and even the entire globe. To confront these transnational problems, interest groups have opted for the transnational approach to deal with these issues. So, it is a matter of fit for

[1]Globalisation and internationalisation are two distinct processes. Internationalisation denotes 'a process of intensifying connections between national domains'. The two activities are summarised as follows: 'the international realm is a patchwork of bordered countries, while the global sphere is a web of transborder networks' (Scholte 1999: 15).

purpose. In effect, the degree and nature of the problem influence the degree and nature of the proposed solution. Stated differently, the form of the problem follows the function of the interest group to tackle the problem.

Thus, in any event, the actions of interest groups *vis-à-vis* a number of institutions and governments included are becoming more transnational in magnitude (Ghils 1992). For instance, in 1992, Wolfgang Pricher[2] gave the following warning to the British Dam Society (BDS). He cogently notes that 'a serious general countermovement…has already succeeded in reducing the prestige of dam engineering in the public eye, and it is starting to make work difficult for our profession' (Pricher 1992). The 'countermovement' Pricher was referring to, consists of a number of interest groups, operating not only within state borders, but also transnationally. That being the case, interest groups have 'global policy goals' to conciliate (Payne 1996: 171). Thus, as one observer aptly puts it, '[i]n contrast to "interstate relations" taking place between the governments of two or more states, transnational interactions involve at least one party that is not governmental in character. In the process, the government of at least one of the interacting countries is bypassed' (Soroos 1986: 13–14).

Stated in another way, interest groups do not only forge links with each other but at an increasing rate with other entities as well. Interest groups are energetic in a large variety of activities at the local, national, transnational, and international levels. They perform functions in all the various issue areas and especially in human rights, humanitarian, and environmental domains, to name but a few (Mingst 1995). Notwithstanding the nature of their interaction with other organisations, the transnational character of interest groups takes on a meaningful significance when the levels of interaction between them and other actors in domestic and international affairs are considered. At least three levels are identified: between interest groups themselves, between interest groups and intergovernmental organisations or agencies of these organisations, and between interest groups and governments. At all three levels, interest groups have put a considerable amount of global pressure on a wide range of issues and platforms by way of transnational campaigns and lobbying (Krut 1997).

Towards a Different Perspective on Interest Groups in Water Politics and Governance

Water resources management, in South Africa, in general, and the Orange-Senqu River, in particular, is consequently a tall order, with climatological, hydrological, and ethical considerations intersecting to make water resources management in this context quite a unique experience not only from an institutional, but also political, scientific, and ethical point of view. A concept or idea that plays an important role

[2]Former president of the International Commission on Large Dams (ICOLD).

in these dimensions is the water–energy–food security nexus. The concept has become somewhat 'voguish'[3] since researchers link the concept to consideration to meet increasing demand for these resources in a transparent and equitable manner (Lele et al. 2013). Researchers also point to the interdependent relationship between the three resources; if, for instance, the price of energy increases so too will the price of food rise (Gulati et al. 2013). The Lesotho Highlands Water Project is a manifestation of the water–energy–food nexus. Not only is the Project delivering more water to an expanding population and economic base in South Africa's Gauteng Province, it is also generating much needed electricity for Lesotho. From an ethical perspective, many interest groups have voiced their concern that the displaced people will be worse off because of the project, since the reservoirs have flooded pastures and agricultural land.

What is more, from a research perspective, the water research community has a particular outlook on the problems facing South Africa, in terms of the above-mentioned aspects. The biophysical profile of the river basin highlights the natural variables influencing water resources management as well as the strategic importance of water resources in the Lesotho and South African economies. It also gives a first impression of the actors involved in its management: states. It is not wrong to device such a profile and initial description of the Orange-Senqu River basin. Hydrology, for instance, is an important scientific discipline that is advancing human understanding of surface and subsurface water resources to such an extent that irrigation schemes and their impact on food production would not have been possible. Said differently, it would not be possible to debate the water–energy–food nexus without hydrology to guide us. Describing a river basin as consisting predominantly of states has the potential to instil a predominantly positivist paradigm, with the potential of placing more emphasis on biophysical processes and state actors in the political economy of a river system and the security considerations accompanying such economic aspects (e.g. Meissner 2013a, b, 2014a, b, c). This could blind us to the finer nuances of the individual Lesotho Highlanders' water–energy–food nexus. Said differently, the water–energy–food security nexus is not only important from a macropolitical and economic perspective, but also from a human or individual security perspective. From a rational cost-benefit analysis, it is important to look at the river basin from such a perspective. It is also equally important to consider the human security needs of the Lesotho Highlanders, something rational choice would find difficult to explain because of its cost-benefit calculations.

A dominant positivist paradigm drives water resources management and the research thereof in South Africa (Meissner et al. 2012), especially when looking at water supply influenced by strategic considerations and biophysical elements. It is possible that the social or political scientists mentioned above, myself included, tried to emulate their colleagues from engineering, hydrology, and botany, such as Conley (1995, 1996), Basson et al. (1997), and Ashton (2000) to map the political and social dynamics of the Orange-Senqu River basin and in so doing try of get as

[3]I would like to thank the anonymous reviewer for this idea.

full a grasp of these dynamics of the river basin as possible. We therefore had a purposive goal in mind, to explain, albeit in a positivist manner, to the uninformed decision-maker that it is not only biophysical aspects that matter, but also political. Alas we have succeeded and failed. Political scientists have succeeded to explain why states do what they do (e.g. the signing of treaties) and how they interact with one another outside the realm of their treaties and the institutions they create and the nature and extent of the strategic importance of a river, like the Orange–Senqu, in the political economy of South Africa, its neighbours and the Southern African Development Community (SADC). Be that as it may, we have also failed—we have not come out very strongly on the part of an alternative paradigm such as post-positivism and focussing on non-state actors as well as the framing of our questions in a different manner. Our questions were always framed in the ambit of the state as the 'Supreme Being' on the international stage and even the river basin. We cannot wish the state and its role in water politics and governance away, and this is certainly not my intention. On the contrary, the state must be kept inside our sphere of analysis, but our organising questions need to be different if we want to progress in our understanding of social phenomena in a natural scientific setting. If not, we will fail further and be kept from gaining in our understanding of agents in water politics and governance.

It is here where Hobson and Seabrooke's (2007) notion of everyday international political economy (EIPE) and Katzenstein and Okawara's (2001/2002) and Sil and Katzenstein's (2010) idea of 'analytical eclecticism' start to play an important role.

If we should frame questions regarding governance, politics, and management outside the ambit of the dominant theoretical stances mentioned above, it will reveal 'information at the local and transnational levels that tell us how the actions of the key players are contested [or supplemented] by everyday actions' (Hobson and Seabrooke 2007: 10). It could also install a mode of thinking whereby different views from different theories and paradigms are amalgamated (Sil and Katzenstein 2010). In other words, it will broaden the research domain and we will 'discover information about how everyday actions inform the dominant processes of [governance, politics, understanding and management]', highlighting transformative moments and processes. This will sketch a more complete vista and move away from distortions (Hobson and Seabrooke 2007: 6) of how governance and politics operate in a multi-varied and complex social and natural environment. It is here where Hobson and Seabrooke's (2007) notion of everyday international political economy juxtaposed with regulatory international political economy becomes appropriate. Table 1.1 summarises these.

The table is informative since it indicates a progressing of International Relations theory, which is applicable to governance and politics within the domestic and transnational domains (see Meissner 2001a, b, 2004a, b, c, 2012; Meissner and Jacobs 2014; Meissner and Ramasar 2014). This is an important observation. As mentioned earlier, progressive development of theory in any scientific discipline is necessary not only to broaden our understanding of reality but also to inform research agendas. In effect, theoretical progression begets policy progression and vice versa.

Table 1.1 Aims and approaches in regulatory and everyday IPE

	Regulatory IPE	Regulatory IPE	Everyday IPE
Organising question	Who governs?	Who benefits?	Who acts and how do their actions enable change?
Epistemology	Positivist/rationalist or interpretivist	Positivist/rationalist	Post-positivist/interpretivist and rationalist
Theoretical tradition	Neorealism/neoliberalism/systemic constructivism	Classical structuralism	Sociological/complexity/social constructivism
Unit of analysis	Great powers (e.g. USA), other states, international regimes, and ideational entrepreneurs	Capitalist world economy, structures of rule	Everyday actors interacting with elites and structures
Prime empirical focus	Supply of order and welfare maximisation by leaders	Maintenance of the powerful and the unequal distribution of benefits	Social transformative and regulatory processes enacted or informed, by everyday actions
Locus of agency	Top-down	Top-down	Bottom-up

Source Hobson and Seabrooke (2007: 6)

The rationale behind analytic eclecticism is to avoid paradigmatic compartmentalisation since this leads to a disjunction between the researcher and what he or she can offer the decision-maker. Arguing from a particular paradigm can become an obstacle of understanding even if it gives powerful insights. Analytic eclecticism does not discard established paradigms or traditions instead 'it explores substantive relationships and revealing hidden connections among elements of seemingly incommensurable paradigm-bound theories, with an eye of generating novel insights that bear on policy debates and practical dilemmas. This requires an alternative way of thinking about the relationships among assumptions, concepts, theories, the organization of research, and real world problems' (Sil and Katzenstein 2010: 2).

One argument would be to put forward complexity theory as the all-encompassing theory that explains all or at least many aspects found in nature and society. Even so, such an argument is bound to run into trouble because of the mere fact that complexity is considered superior or parsimonious for explaining and solving all problems. As such, it contains a high degree of error when relying on only one theoretical lens (Sil and Katzenstein 2010). Sil and Katzenstein (2010: 2) maintain that '…we can and should do a better job or recognizing and delineating relationships between concepts, observations, and causal stories originally constructed in different analytic perspectives'. This means that complexity would be limited in scope for it would be only one analytic perspective. A researcher, arguing from a complexity stance would 'miss' other perspectives because of the perception that complexity is all encompassing in explaining issues and phenomenon. What is more, complexity might be useful in explaining many elements at work in social-ecological systems, but it has little to offer decision-makers interested in bringing about change in society. Complexity is not a reliable blueprint and can desensitise us to the possibility that the theory is wrong (Hirschman 1970; Tetlock 2005 in Sil and Katzenstein 2010). Furthermore, complexity is a theory and not a paradigm.

Analytical eclecticism tackles the issue of paradigmatic eclecticism at the ontological and epistemological levels. What divides paradigms are not the substantive claims about phenomena but their metatheoretical assumptions regarding 'how such claims should be developed and supported' (Sil and Katzenstein 2010: 4). It does not slice up complex social phenomena just for making them simple and easy to analyse. In other words, reductionism is not an underlying premise. Important substantive questions with relevant real-world application are in the offing by integrating empirical observations and causal stories. This brings about the 'promise of richer explanations' (Sil and Katzenstein 2010: 3). In other words, it facilitates the quantum leap from singular explanations of real-world problems to fuller explanations, alternatives, and solutions to such problems. Where paradigms have blind spots, they have, at the same time, useful insights into issues, challenges, and opportunities. There are therefore connections and complementarities between paradigms to exploit. This could lead to a situation where more useful theoretical and empirical insights could be generated (Sil and Katzenstein 2010).

Conclusion

The conceptualisation of non-state entities, as *interest group*, *pressure group*, *lobby,* and *NGO/INGO*, has a long history. Many scholars studying these entities usually define them within the subject matter they are examining. No matter what, I show that there are a number of definitions in Political Science and International Relations to describe them. The different conceptualisations of the terms *interest group*, *pressure group*, *lobby,* and *NGO/INGO* have caused considerable confusion within Political Sciences on what exactly an interest group is. In this chapter, it was shown that to argue over different definitions of these entities is not important. What is of essence is their role and activities in society. Of special importance is their relationship with the state, government, and international institutions, and their impact on state autonomy, and the governmental policy process. Following from interest groups' relationship with governments and other private sector actors, analytic eclecticism and the framing of problems from a non-state centric perspective, holds the promise of a more nuanced understanding. The study of interest group involvement in the Lesotho Highlands Water Project is an attempt at giving credence to the idea and practice of analytic eclecticism.

References

Archer, C. 1992. *International organizations*, 2nd ed. London and New York: Routledge.

Ashton, P. 2000. *Southern Africa water conflicts: Are they inevitable or are they preventable?*. Geneva: Green Cross International.

Basson, M.S., P.H. Van Niekerk, and J.A. Van Rooyen. 1997. *Overview of water resources availability and utilization in South Africa*. Pretoria: Department of Water Affairs and Forestry.

Baumgartner, F.R., and B.L. Leech. 1998. *Basic interests: The importance of groups in politics and in political Science*. Princeton, New Jersey: Princeton University Press.

Conley, A.H. 1995. A synoptic view of water resources in Southern Africa. Paper presented at the conference of Southern Africa foundation for economic research on integrated development of regional water resources, in Nyanga, Zimbabwe.

Conley, A.H. 1996. The need to develop the water resources of Southern Africa. Paper Presented at the Vic Falls conference on aquatic systems, July 1996, in Zimbabwe.

Geldenhuys, D. 1998. *Foreign political engagement: Remaking states in the post-Cold War world*. Great Britain and the United States: Macmillan Press and St. Martin's Press.

Ghils, P. 1992. International civil society: International non-governmental organizations in the international system. *International Social Science Journal* 44(131–134): 417–431.

Gleick, P.H. 1998. *The world's water 1998–1999: The biennial report on freshwater resources*. Washington, D.C.: Island Press.

Grant, W.P. 2000. *Pressure groups and British politics*. London and New York: Macmillan Press and St Martin's Press.

Gulati, M., I. Jacobs, A. Jooste, D. Naidoo, and S. Fakir. 2013. The water-energy-food security nexus: Challenges and opportunities for food security in South Africa. *Aquatic Procedia* 1(1): 150–164.

Heywood, A. 1997. *Politics*. London: Macmillan Press.

Hirschman, A.O. 1970. The search for paradigms as a hindrance to understanding. *World Politics* 22(3): 329–343.

Hobson, J.M., and L. Seabrooke. 2007. Everyday IPE: Revealing everyday forms of change in the world economy. In *Everyday politics of the world economy*, ed. J.M. Hobson, and L. Seabrooke. Cambridge: Cambridge University Press.

Holsti, K.J. 1995. *International politics: A framework for analysis*. Englewood Cliffs, N.J.: Prentice Hall International, Inc.

Katzenstein, P.J. and Okawara, N. 2001/2002. Japan, Asian-Pacific security, and the case for analytical eclecticism. *International Security*, 26(3): 153–185.

Krut, R. 1997. *Globalization and civil society: NGO influence in international decision-making*. Geneva: United Nations Research Institute.

Lele, U., M. Klousia-Marquis, and S. Goswani. 2013. Good governance for food water and energy security. *Aquatic Procedia* 1(1): 44–62.

Mackenzie, W.J.M. 1955. Pressure groups in British government. *The British Journal of Sociology* 6(2): 133–148.

McCully, P. 1996. *Silenced rivers: The ecology and politics of large dams*. London: Zed Books.

Meissner, R. 2001a. Interest groups impacting on the agential power of the state in international river basins: Towards a framework for analysis. Paper presented at the South African Association of political studies, October 5–7, 2001, in Durban.

Meissner, R. 2001b. Water disputes: Drawing the line. *Conflict Trends* 2(2001): 34–37. Durban: ACCORD.

Meissner, R. 2004a. *The transnational role and involvement of interest groups in water politics: a comparative analysis of selected Southern Africa case studies*. D.Phil Dissertation: University of Pretoria, Faculty of Humanities.

Meissner, R. 2004b. Foreign policy no longer the domain of governments only. *Foreign Policy Monitor* 2004: 1–2.

Meissner, R. 2004c. Interest groups give people a voice. Business Day, 12 October 2004: 11.

Meissner, R. 2013a. *Changing theoretical lenses broadens research space*. Pretoria: Council for Scientific and Industrial Research. Report no. CSIR/NRE/WR/MEMO/2013/0006/A.

Meissner, R. 2013b. *PULSE³: A Governance assessment framework and its application to selected waste water treatment works in the Greater Sekhukhune District Municipality*. Pretoria, South Africa: Council for Scientific and Industrial Research.

Meissner, R. 2014a. A critical analysis of research paradigms in a subset of marine and maritime scholarly thought. In *Reflections on the state if research and technology in South Africa's marine and maritime sectors*, ed. N. Funke, M. Claassen, R. Meissner, and K. Nortje. Pretoria: Council for Scientific and Industrial Research.

Meissner, R. 2014b. *Paradigms and theories influencing a selected section of global water governance: The case of the UNDP's water and ocean governance focus area*. Pretoria: Council for Scientific and Industrial Research. Report no. CSIR/NRE/WR/IR/2014/0077/A.

Meissner, R. 2014c. *Paradigms and theories in water governance: The case of South Africa's national water resource strategy*, 2nd ed. Pretoria: Council for Scientific and Industrial Research. Report no. CSIR/NRE/WR/IR/2014/0076/A.

Meissner, R., and I. Jacobs. 2014. Theorising complex water governance in Africa: the case of the Proposed Epupa Dam on the Kunene River. *International Environmental Agreements: Politics, Law and Economics* 14(2): 1–28.

Meissner, R., and V. Ramasar. 2014. Governance and politics in the upper Limpopo River basin. *South Africa. GeoJournal* 79(5): 1–21.

Meissner, R., Claassen, M., Funke, N., Ntombela, C. 2012. Proceedings of workshops focusing on various aspects of South African water politics, governance and institutional arrangements. Pretoria: Council for Scientific and Industrial Research. Publication number: CSIR/NRE/ECOS/IR/2012/0010/C.

Mingst, K.A. 1995. Uncovering the missing links: Linkage actors and their strategies in foreign policy analysis. In *Foreign policy analysis: Continuity and change in its second generation*, ed. L. Neack, J.A.K. Hey, and P.J. Haney. Englewood Cliffs, New Jersey: Prentice Hall.

Moulder, S., H. Scott, and H. Van der Merwe. 1982. *Survey of interest groups and groups concerned with conflict*. Cape Town: Centre for Intergroup Studies, University of Cape Town.

Nye, J.S., and R. Keohane. 1971. Transnational relations and world politics: An introduction. *International Organization* 25: 329–349.

Payne, R.A. 1996. Non-profit environmental organizations in world politics: Domestic structure and transnational relations. *Policy Studies Review*, 14 (Spring/Summer 1995): 171–182.

Petracca, M.P. 1992. The rediscovery of interest group politics. In *The politics of interests: Interest groups transformed*, ed. M.P. Petracca. Boulder, Colorado: Westview Press.

Pricher, W. 1992. 36 000 large dams and still more needed. Paper presented at the Seventh Biennial Conference of the British Dam Society, University of Stirling, June 25, 1992.

Soroos, M.S. 1986. *Beyond sovereignty: The challenge of global policy*. Columbia, South Carolina: University of South Carolina Press.

Scholte, J.A. 1999. The globalization of world politics. In *The globalization of world politics: An introduction to international relations*, ed. J. Baylis, and S. Smith. Oxford: Oxford University Press.

Sil, R., and P.J. Katzenstein. 2010. *Beyond paradigms: analytic eclecticism in the study of world politics*. Houndmills, Basingstoke: Palgrave Macmillan.

Smith, M.J. 1993. *Pressure, power and policy: State autonomy and policy networks in Britain and the United States*. Hertfordshire, Great Britain: Harvester Wheatsheaf.

Stern, G. 2000. *The structure of international society: An introduction to the study of international relations*. London: Pinter.

Tetlock, P. 2005. *Expert political judgement: How good is it? How can we know?*. Princeton, N.J.: Princeton University Press.

Truman, D.B. 1951. *The governmental process: political interests and public opinion*. New York: Alfred A. Knopf, Inc.

Turton, A.R. and Ohlsson, L. 1999. Water scarcity and social adaptive capacity: Towards an understanding of the social dynamics of managing water scarcity in developing countries. Paper presented in the workshop no. 4: Water and social stability of the 9th Stockholm water symposium "Urban stability through integrated water-related management", hosted on 9–12 August by the Stockholm Water Institute (SIWI) in Stockholm, Sweden. Also available as MEWREW occasional paper no. 18.

Turton, A.R. 2000b. Precipitation, people, pipelines and power: Towards a 'virtual water' based political ecology discourse. MEWREW Occasional Paper No. 11. http://www2.soas.ac.uk/Geography/Waterissues/Home.html, June 26, 2002.

Willets, P. 1997. Transnational actors and international organizations in global politics. In *The globalization of world politics: An introduction to international relations*, ed. J. Baylis, and S. Smith. Oxford: Oxford University Press.

Wilson, G.K. 1990. *Interest groups*. Oxford: Basil Blackwell, Ltd.

Chapter 3
The Orange-Senqu River

Abstract In this chapter, I outline the biophysical characteristics of the Orange-Senqu River. I also investigate the importance of the river to the Lesotho and South African economies. A large part of the chapter is devoted to an analysis of the historical development of water resources management in the river basin. This is an important consideration, since the hydropolitical history of a river basin indicates which actors were or are dominant in the governance and politics of a river basin and water projects implemented to harness the rivers' potential.

Keywords Orange-Senqu River · Irrigation · Agriculture · Lesotho · South Africa · Lesotho Highlands Water Project Treaty

Introduction

The Orange River is probably South Africa's most important surface water source and hence also the country's, and arguably Southern Africa's, most developed river. The river supplies South Africa's economic heartland, Gauteng Province, with water through a series of elaborate inter-basin transfer schemes. This chapter briefly explores the Orange-Senqu River's biophysical characteristics in terms of its hydrological and geographical boundaries as well as the river's importance. The biophysical characteristics of the river and its importance to the countries through which it flows are intertwined.

Biophysical Characteristics

With its source deep in the Maluti Mountains of Lesotho, the Orange-Senqu River is an international river with four countries sharing its basin: Botswana, Namibia, Lesotho, and South Africa (see Fig. 3.1). The Orange-Senqu rises some 3300 m above

R. Meissner, *Interest Groups, Water Politics and Governance*,
SpringerBriefs in Environmental Science, DOI 10.1007/978-3-319-21130-5_3

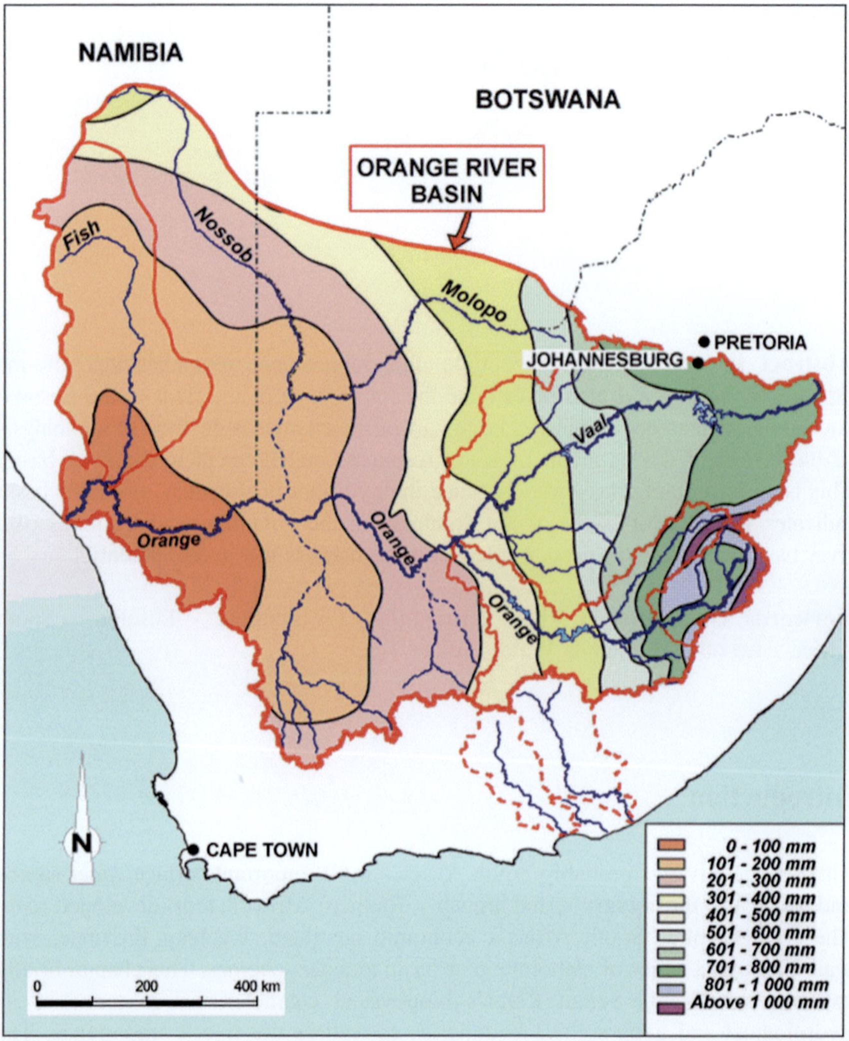

Fig. 3.1 The Orange River basin indicating the various inter-basin transfers (*Source* DWA 2013)

sea level in the region of *Mont-aux-Sources*. Here, in the Lesotho Highlands, the Orange is known as the Senqu River. After flowing out of Lesotho, the Orange streams through South Africa until it forms the border between Namibia and South Africa for 450 km before emptying in the Atlantic Ocean. The river has a total length of about 2300 km with a catchment area of around 964,000 km^2. The Vaal River, situated entirely in South Africa, is the Orange's most important tributary and supplies most of South Africa's economic heartland—Gauteng Province—with water. Of the Orange River's total catchment, 4 % is situated in Lesotho, 62 % in South Africa, 9 % in Botswana, and 25 % in Namibia (McKenzie and Roth 1994;

Conley 1995, 1996; Heyns 2003; Turton 2003a, b; DWAF 2009; Jacobs 2012). Of the total mean annual run-off, each riparian country makes an unequal contribution to the Orange River's flow. South Africa contributes 55 %, Botswana 0 %, Lesotho 41 %, and Namibia 4 %. Botswana contributes 0 %, because the Molopo and Nossob rivers do not contribute any water to the flow of the Orange River (Heyns 2003). Even so, groundwater might be contributed from Botswana's portion of the basin to the flow of the river (Jacobs 2012), but since the volume is not known, it is considered that it contributes no water to the Orange-Senqu River basin.

The Orange River has a virtual mean annual run-off (MAR) of between 10.6 and 12 billion cubic metres of water per year (bcm/year). This reflects the volume of water within the river system before any development with associated abstraction has taken place. Irrespective of this, the Orange is the most developed river system in Southern Africa. The highest proportion of development takes place in South Africa's portion of the basin (Heyns 2003; Turton 2003a). The Orange River carries about 20 % of South Africa's surface water resources, while its main tributary, the Vaal, provides most of the water required by the Gauteng Province (Basson et al. 1997; Turton 2003a; Meissner 2004), indicating the prominence of surface as well as subsurface or groundwater resources in South Africa.

The Orange River's Importance

The significance of water in South Africa becomes striking when looking at South Africa's climatic conditions, especially rainfall. South Africa is a water-scarce country. On average South Africa receives 450 millimetres (mm) of rainfall per annum, while the global average is in the region of 860 mm per year (DWAF 2008; SA Yearbook 2011). Rainfall is highly variable in terms of geographical distribution as well as time (Schulze 1997; Palmer and Ainslie 2006). The semi-arid climate, naturally limited water resources, and rainfall variability, accentuated by predicted climate change, all pose biophysical constraints for future development (Ashton 2000) on the country's water resources. In addition to this permanent constraint of limited water resources, South Africa also has to address the challenge of addressing historically inequitable water access among its population, as well as ensuring that ecological water requirements are met (Brooks and Wolfe 2007; Pott et al. 2009). This is aggravated by climate change and international obligations to neighbouring countries with shared river basins (Claassen 2010).

Together, the Vaal and Orange rivers are South Africa's most important strategic surface water resources. These rivers supply water for a number of economic activities inside and outside the Orange basin, ranging from agricultural to industrial and from urban to mining activities (Turton 2003a, b; Meissner 2004). According to the latest (2011) census data, almost 24 % of South Africa's population lives in the Gauteng Province. The 2011 census results show that there had been an increase of 1.7 % from 22 to 23.7 % in 2011 in the Province's population. This is the largest growth rate of all the nine provinces. Of all the provinces,

Gauteng is also the most populated, followed by KwaZulu-Natal 19.8 %, Eastern Cape 12.7 %, Western Cape 11.2 %, Limpopo 10.4 %, Mpumalanga 7.8 %, the North-West Province 6.8 %, Free State 5.3 %, and the Northern Cape 2.2 % (Stats SA 2011).

Most, but not all of the urban, industrial and mining activities are situated within the Gauteng Province. Water is supplied to the Province as well as other provinces, such as the platinum mines in the North-West Province and coal-fired power stations in Mpumalanga, through an elaborate arrangement of inter-basin transfer schemes. Since the 1960s, a number of these schemes have been implemented, most notably the Tugela-Vaal River Scheme and the LHWP (Turton 2003a; Meissner 2004). The Orange River, in particular, plays a central role in the network of inter-basin transfer schemes that criss-cross the South African landscape. The importance of inter-basin transfer schemes in the South African economy is illustrated by the proportion of gross geographical product (GGP) that is supported by these transfers in each of the country's nine provinces (Table 3.1).

The table does not only indicate that 100 % of Gauteng's GGP is supported by inter-basin transfer schemes, but also indicate that water needs to be transported from where it is in abundance to where it can be utilised for economic activity in other provinces. This reality was mainly driven by industrial development in the post-Second World War era, combined with constant population increases and an accompanying rise in living standards. These drivers placed new demands on South Africa's water resources and necessitated an urgent and innovative way to provide water where it is needed. This was accomplished through inter-basin transfer schemes (DWA 1987; Turton et al. 2004). The Orange River is classified as being 'in deficit', making further developmental opportunities quite difficult (Conley 1995, 1996; Turton 2005), at least in the southern portion of the river basin. Successive South African governments dealt with this reality through an engineered solution, the so-called hydraulic mission (Reisner 1993; Turton and Ohlsson 1999). This mission can be defined as the rationale that underpins the state's aspiration to bring about conditions that are advantageous to socio-economic and political stability. The hydraulic mission can be seen as an ideology in the study and execution of water

Table 3.1 Province's GGP supported by inter-basin transfer schemes

Province	Percentage of GGP
Eastern Cape	70
Free State	65
Gauteng	100
KwaZulu-Natal	70
Limpopo	30
Mpumalanga	70
Northern Cape	50
North-West	90
Western Cape	70

Sources Basson et al. (1997) and Turton (2003a, b)

politics. As such, the hydraulic mission infuses itself into the dominant discourse or paradigm that legitimises state actions (Reisner 1993; Turton and Ohlsson 1999; Turton et al. 2000; Warner 2000; Warner and Meissner 2008; Meissner and Turton 2003; Meissner 2004), in the construction of inter-basin transfers.

It is, however, not only states that are part of the hydraulic mission. Economic resources are needed to execute such projects since they are quite costly. Usually, an arrangement with a group of financial institutions, such as the World Bank, is necessary. The hydraulic mission is in a sense dominated by the state and its government apparatus.

The Lesotho Highlands Water Project

This section gives an overview of the LHWP in terms of its technical specifications, institutional arrangements, and the benefits both Lesotho and South Africa get from its construction.

The LHWP is an international inter-basin transfer scheme jointly implemented by Lesotho and South Africa, the purpose of which is to divert water from the upper reaches of the Orange-Senqu River in the Lesotho Highlands to Gauteng—supplying water to the Vaal River system. Its other main purpose is to generate hydroelectricity for Lesotho. The joint venture consists of several major and minor dams, a series of water transfer tunnels dug through the Maluti Mountains and various associated infrastructures, including hydroelectric generators and pumping stations. More than 90 % of the project is located in Lesotho. The LHWP consists of a number of completed and planned phases, namely Phases 1 (A and B), 2, 3, and 4 (Gleick 1998; Horta 1996; Meissner 1998; LHDA and TCTA 2001; Meissner and Turton 2003).

Phase 1A of the project is designed to transfer water at a rate of 18 m^3/s and generate 72 MW of electricity and consists of two dams, Katse and Muela, excavation of 82 km of subterranean water transfer tunnels and the construction of an underground hydroelectric plant. The Katse Dam is considered the 'jewel' of the scheme. It is 185 m high and has a reservoir surface area of 35 km^2 and a reservoir capacity of 1.9 billion cubic metres (bcm) (James 1980; DWAF 1994; Wallis 1996; Mochebelele 2000).

Phase 1B consists of the construction of two dams (Mohale [145 m high] and Matsoku) connected to the Katse reservoir. This phase was completed in January 2003. It delivers water at a rate of 12 m^3/s. By 2003, it will transfer an average of 871 mcm of water per year through a network of 260 km of tunnels (DWAF 1994, 1998; Wallis 1996; Hoover 2001; LHDA and TCTA 2001). Phases 2–4 will eventually increase the water transfer capacity to around two billion cubic metres per year (bcm/year) in about 10 years from now. In August 2011, the cooperation agreement for Phase 2 of the project was signed between Lesotho and South Africa with the Phase currently under construction. Phase 2 consists of the construction of the Polihali Dam and a water transfer tunnel to Katse Dam. The overall purpose is to further increase the water transfer to the Vaal River system (DWA

2012a). The Kobong pump storage scheme is also part of this phase. This scheme will generate 1200 MW of electricity on completion. The Katse reservoir will be utilised as the lower reservoir, and an upper reservoir will be constructed in the Kobong Valley for electricity generation. Construction using this scheme is envisaged to start in March 2015, and completion is set for 2019. Lesotho is seeking customers such as Eskom and other large users of electricity to supply electricity. The cost of the pump storage scheme is R7.6 billion and will be financed by Lesotho. The total cost of Phase 2 is in the region of R15.4 billion (DWA 2012b; TCTA 2013).

Apart from direct benefits, there are also a number of indirect benefits both Lesotho and South Africa will gain from the scheme's implementation. These include stimulation and acceleration of socio-economic development of the Lesotho Highlands; sharing of financial savings by not implementing the historically mooted Caledon Cascades Scheme; the minimisation of air pollution generated by coal-fired power stations to pump water through the Cascades Scheme; fostering of economic and political independence in both countries; job creation during the construction phases; unpolluted water from Lesotho will increase the quality of that of the Vaal River system upstream from the Vaal Dam; and status and prestige because of the size of the project and some of the engineering feats to both countries, but more so to Lesotho due to its small land size (Treaty 1986; Du Pisani 1992; Conley and van Niekerk 1998; Meissner 1998; Hoover 2001; Turton 2003b).

The description of the LHWP in terms of its specifications regarding delivery of water, generation of electricity, different implementation phases, and secondary or indirect benefits to both implementing countries would fit neatly within the ambit of the 'benefit sharing' (Sadoff and Gray 2002; Qaddumi 2008; Turton 2008) discourse on an interstate level as well as the fostering of interstate cooperation and the stimulation and promotion of economic regional integration.

The project is managed by the Lesotho Highlands Development Authority (LHDA), charged with the implementation, operations, and maintenance thereof in Lesotho. It is also responsible for construction, environmental protection, and all resettlement and compensation issues. In South Africa, the Department of Water and Sanitation (DWS) and the Trans-Caledon Tunnel Authority (TCTA) oversee the project (DWA 2012). The TCTA has the same responsibility as the LHDA, but only for those components to be implemented, operated, and maintained in South Africa. The Lesotho Highlands Water Commission (LHWC), formerly the Joint Permanent Technical Commission (JPTC), was established in 1986 to represent both countries. The LHWC has monitoring and advisory powers over the administrative, technical, and financial activities (Gleick 1998; Meissner 2000a, b, c; LHWP 2003; TCTA 2003; Turton 2003b). There is the potential for boredom (e.g. Rosenau 1990) to set in if analyses do not move beyond the state-centric conceptualisation to include other actors and structures to the investigation of the LHWP. Such an investigation will start by looking at the hydropolitical history of the Orange River.

Historical Background of Water Resources Development in the Orange River Basin

The hydropolitical history of the Orange River can be divided into a number of phases. The first of these deals shortly with the period 1867–1960, when water resources management projects started to appear in the Orange River basins. The second phase deals with the period 1956–1986. Accordingly, a third phase is identified from 1986 to 2013.

Water Project Implementation Picks up: 1867–1956

Diamonds were discovered on the banks of the Orange River in the Hopetown District in 1867 and later at Kimberley in 1870. This was followed by the discovery of gold and the subsequent establishment of Johannesburg in 1886 in the Witwatersrand area. These events were significant for the utilisation of the Orange and Vaal rivers' water resources in that the discovery of the minerals stimulated migration into the interior of South Africa from the coastal areas. This led to the subsequent establishment of towns and markets and agriculturalists started to farm more intensely, which necessitated irrigation (Thompson and Lamar 1981; Bath 1999; Turton et al. 2004). The discovery of diamonds and gold in the Orange River basin are trigger events, resulting in South Africa's hydraulic mission (Turton and Meissner 2000; Reisner 1993). It was not as if the hydraulic mission started on a specific date by a specific entity. It happened in fits and starts and took on a gradual character. In the Orange River basin, as far as historical records are concerned, it was an individual that saw the potential of irrigation and implemented a project to harness this irrigation potential.

It is thought that the Dutch Reformed Church got the first irrigation system going at Kakamas in 1883. Under the guidance of Reverend Christian Schröder, an irrigation furrow was completed on 2 May 1885 and the missionary station was able to irrigate its gardens from the Orange River (Macdonald 1913; Green 1948; Hopkins 1978; Turton et al. 2004; Van Vuuren 2012). According to Legassick (1996: 371–372): 'In organizing [the Upington canal's construction], Schroeder…took [his] lead from Abraham September [a farmer], who had first led water from the Orange river'. It would appear that the Kakamas works were constructed after the Upington irrigation canal (Meissner 2014). According to the *Standard Encyclopaedia of South Africa* (1975 cited in Legassick 1996: 372), 'Upington owes its prosperity mainly to agriculture and the development of irrigation along the Orange River. Here, at Upington, Schroeder as missionary among families of mixed European and other blood designed the first irrigation canal of the lower Orange River, a scheme so successfully applied at Kakamas in later years'. Legassick's (1996) research does not only indicate that previous historians were incorrect, but also incorrect in a profound manner. Abraham September, as a non-white landowner and farmer, was most likely responsible for the first irrigation works on the Orange River. What is more, an institution, the Dutch Reformed Church, took its cue from September showing how a change-enabling agent can influence a collectivity like the church.

From then onwards, the management of the Orange River's water resources took on a different tone and government stepped into get large irrigation works off the ground. After the establishment of the Union of South Africa, in 1910, A.D. Lewis, of the Department of Irrigation (the forerunner of the Department of Water and Sanitation), did an extensive reconnaissance of the lower Orange River basin for the construction of more water projects (Conley and van Niekerk 1998; Turton 2003a). Eighteen years later, Lewis proposed the development of a tunnel to transfer water from the Orange River to the drought-prone Eastern Cape. This plan was the harbinger of the Orange River Project that was implemented in the 1960s. The plan had already been mooted by Sir George Grey in the 1850s and later by Thomas Bain in 1886 (Turton 2003a; Meissner 2004a; Turton et al. 2004).

South Africa suffered a major drought from 1929 to 1931 and was in the grips of the Great Depression. After these events, the Department of Irrigation, on instruction by the Prime Minister, launched a number of nationwide poverty-relief programmes. The main aim was to supply employment to so-called poor whites, who had been impoverished during the Great Depression of the 1930s, and to implement irrigation projects to get agriculture on a sound footing. These events and the reaction they invoked regarding the construction of water management projects can be seen as the second fundamental component of the hydraulic mission in the Orange River basin. During this phase, large-scale labour-intensive irrigation and other projects were implemented on the Orange and Vaal rivers. The most notable of these is the Orange River Project of the 1960s (Conley and van Niekerk 1998; Turton 2003a; Meissner 2004; Turton et al. 2004).

From Idea to Treaty: 1956–1986

The Orange River Project was South Africa's first inter-basin transfer scheme, with construction starting in the late 1960s. Other inter-basin transfer schemes were also implemented after the Orange River Project, most notably the Tugela-Vaal Transfer Scheme (RSA 1962; Conley and van Niekerk 1998; Turton 2003a; Meissner 2004; Van Vuuren 2012) supplying water to the Vaal River system. Before the implementation of all these projects, the Lesotho Highlands Water Project was investigated by consulting engineers in the 1950s.

A New Idea and Political Issues Surfacing

The Lesotho Highlands Water Project originated from two main considerations, namely that the Vaal Dam, constructed in the 1930s, will not be able to meet the growing water needs indefinitely and that the Lesotho Highlands are a potentially reliable source of water supply. In 1950, the High Commissioner to Lesotho, Sir Evelyn Baring requested a survey of the water potential of the country. He realised that water was the only natural resource Lesotho had in abundance and therefore gave the country some advantage in its economic development with its powerful neighbour, South Africa. Sir Peter Ballenden, director of the Department of Public Works, chose the engineer Ninham Shand to determine the viability of exporting Lesotho's water (Brooks 1970; Van Robbroeck 1986; LHDA and TCTA 2001; Van Vuuren 2012). After a reconnaissance mission into the Maluti Mountains, Shand published a plan to harness the upper reaches of the *Senqu* River and transport the water to the Orange Free State gold mines. At that time, the project was known as the Oxbow Scheme. This project involved the construction of a high-altitude dam, a hydroelectric power station, and a tunnel through the mountains. Surprisingly, the plan was rejected, but in the mid-1960s, a drought caused renewed interest (Eksteen 1972; Van Robbroeck 1986; LHDA and TCTA 2001).

The original idea was to sell both water and electricity to South Africa. Lesotho was still, at that time, a poor and underdeveloped protectorate of the United Kingdom, and the influx of foreign revenue from the sales would have been welcomed. It was therefore assumed that Lesotho and South Africa would both benefit from the scheme. South Africa would get water to sustain gold production, and Lesotho much needed development in the form of infrastructure (Smit 1967a: 40, 1967b; Eksteen 1972) and foreign revenue supplementing its treasury. South Africa's participation was, therefore, crucial as will be seen later on.

From its inception, it was realised that the scheme's success would rest entirely on South Africa's willingness to buy water and electricity. In addition, in 1956, the UK announced that South Africa's cooperation was necessary for the construction of the Oxbow Scheme. Even so, South Africa did not give this guarantee. Because Lesotho was less developed than South Africa, and it was impossible for the country to solely implement a project, with a cost of R24 million, it was necessary for South Africa to become a partner in the venture (Eksteen 1972) because of the economic resources South Africa had in hand compared to Lesotho.

South Africa's unwillingness to be a partner changed during the period 1966–1967. In that year, the country's water managers calculated and realised that the Vaal Dam would be insufficient to provide water to the country's economic heartland. The period 1966–1967 was also a period of drought. Consequently, in 1966, the South African government set up a Commission to look into the matter of alternative water resources and hydroelectricity from Lesotho. The Commission concluded that it would be in South Africa's best interest to implement the project, but that the country should not be dependent on water and electricity from Lesotho. The scheme would only serve as a supplementary source of water (White 1965;

Eksteen 1972). A water scarcity therefore prompted the South African government to look for alternative water sources. Yet, South Africa was still not willing to go into a joint venture with Lesotho because of the role that security and politics played. South Africa's interest was defined in terms of Lesotho being a supplementary source of water and electricity and not an outright high-volume supplier.

A number of political issues were identified that could jeopardise the scheme: (1) South Africa's insistence on the incorporation of Basutoland into the Republic, (2) South Africa's *apartheid* policy and the critique it attracted; (3) the Basutoland Congress Party's (BCP) demand that territory, lost in the Basotho Wars of the nineteenth century, be handed back; and (4) South Africa's unwillingness to be dependent on a foreign state for its water and electricity needs (Shand 1956; Young 1961; Eksteen 1972; Barber and Barratt 1990). Lesotho's willingness to go into a partnership with South Africa became a card the Mountain Kingdom played in asserting its own interests and identity. Mere size of South Africa's economy, military might, and land and population size would not make a difference.

Nonetheless, in March 1967, a preliminary feasibility study was presented to the Lesotho government. This study was conducted by Ninham Shand and partners in association with Merz and McLellan. Discussions of the report's proposal with the South African authorities resulted in substantial changes to the design of the project (Van Robbroeck 1986; Turton 2003a; Turton et al. 2004; Meissner 2004). In the following decades, political issues, particularly *apartheid*, would have a significant influence on the interaction between Lesotho and South Africa concerning the Lesotho Highlands Water Project. These issues also defined South Africa's security concerns in the Southern African region, in general, and over the project, in particular.

The question of South Africa's *apartheid* policy was, before Lesotho's independence, already a thorn in Lesotho's side. The then Minister of Economic Development, Molapo, stated that if South Africa should buy water and electricity from the Oxbow Scheme, it would change Lesotho's economy drastically. Despite this, he also declared that Lesotho was afraid that if South Africa got a hold on his country's economy, it would impose its *apartheid* policy on it (Eksteen 1972). Something the Mountain Kingdom resisted vehemently.

Independence and Cooperation

After Lesotho gained independence in October 1966, Premier Lebaua Jonathan announced that the Oxbow Scheme was high on his country's development list. In fact, Jonathan stated that if the scheme were to be implemented, Lesotho's budget would show a positive balance for the first time. Subsequently, negotiations between Lesotho and South Africa took place, and on 23 February 1968, Jonathan announced that an agreement in principle was reached. Both countries hailed this step as the beginning of a long-term positive relationship (Smit 1967b; Eksteen 1972). Construction of the scheme did not start immediately, because

South Africa was implementing the Tugela-Vaal Transfer Scheme and South Africa was not ready to be dependent on water and electricity from an 'unreliable state' (Barber and Barratt 1990). Already the hydraulic mission was hampered by the political circumstances rather than insufficient technical and human resources.

For instance, negotiations on the implementation of the 1966 proposal between Lesotho, assisted by the World Bank and South Africa, failed to produce an agreement and were terminated at the end of 1972. Van Robbroeck (1986: 2) writes: 'The reason for this failure was the vastly different perceptions the two countries had on the Royalty to be paid for the water. The [Republic of South Africa] RSA originally (in 1968) offered a tantieme of 0.5 c/m^3, which was later raised to 1.25 c/m^3, over and above the full cost of water production. Lesotho on the other hand, as advised by the World Bank, wanted a return of [8 %] per annum on capital invested. The RSA argued that this was unreasonable, because Lesotho did not put up equity, but relied on loans, which were fully serviced by the RSA'.

In addition, South Africa could increase the capacity of the Tugela-Vaal at a much lower capital cost. This was due to the provisions made for extensions in the first phase, in the form of the Sterkfontein Dam, that would meet the water demands of the Vaal consumers until 1992 (Van Robbroeck 1986). Nonetheless, future political relations would have a greater impact on the Lesotho Highlands Water Project than engineering solutions.

Worsening Relations

In the beginning to mid-1970s, relations between South Africa and Lesotho started worsening so much so that in 1975, the South African government classified Lesotho 'an extremist state'. Despite this, negotiations on the LHWP were reopened, but because of the 1976 Soweto uprisings in South Africa and the subsequent violent repression of the unrest, talks were suspended. Between 1976 and 1978, the project came to a virtual halt with South Africa unwilling to pay the full price of the water. It wanted a 50 % discount, but Iran[1] convinced it to pay the price Lesotho was asking and the dispute ended (***African Research Bulletin***, 15 July–14 August 1977; Wilsenach 1982; Van Robbroeck 1986; Barber and Barratt 1990; Meissner 1998; Turton 2003b; Meissner 2004). The low-key involvement of the then Iranian government indicates that the Lesotho Highlands Water Project had 'gone transnational', so to speak, albeit at a state level.

In 1978, the LHWP was revived, but the two governments still disagreed on a few issues. The project was revived when the Planning Division of the Department of Water Affairs produced an internal report, recommending that the Upper Orange be considered a source of water for the Vaal River. A larger-scale development

---

[1]Before the 1979 Iranian Revolution, Iran was one of the potential financiers of the LHWP and one of the South Africa's largest trading partners (African Research Bulletin, 15 July–14 August 1977).

project was now feasible. This was after consideration given to the exponential nature of the demand growth for water and the time elapsed since the Oxbow Scheme was advanced. At this stage, analysts pointed to the economic interdependence between the countries, regarding the project (Wilsenach 1982; Van Robbroeck 1986; Turton 2003b). The discourse had changed from one of dependence to one of interdependence between the two countries.

Henry Olivier and Associates, consulting engineers, were appointed to carry out some desk studies, following which discussions with Lesotho were reopened. After the political events of 1976, it was agreed that a joint preliminary feasibility investigation be launched. Each country was to appoint its own consultants, under the direction of a Joint Technical Committee (JTC), which held its first meeting in 1978 (Van Robbroeck 1986). Henry Olivier and Associates and Binnie and Partners were instructed to collaborate in the production of a joint preliminary feasibility report. Lesotho insisted on two conditions, namely that all layouts considered were to include hydroelectric power development in Lesotho itself and that no layouts were to involve storage capacities on the Caledon River (Van Robbroeck 1986). This was one of the first concrete indications that Lesotho wants to utilise the project for political advantage as well. All the same, these conditions had an important impact on the outcome of the study. In May 1979, the JTC produced a report on the strength of which it was decided to proceed with a final feasibility study. Each country was to contribute half the cost to the study (Van Robbroeck 1986). Cooperation was unfortunately inhibited due to the conflictual situation that prevailed between the countries.

The Feasibility Study

According to Van Robbroeck (1986), 'It took Lesotho a considerable time to mobilise funds for its share of the cost of the feasibility study. Although it had provisionally been agreed to establish a joint body to appoint joint consultants, conditions attached to the funding acquired [sic] by Lesotho from the European Development Fund prevented this, and a complicated arrangement had to be devised for co-ordinating and supervising the study.' Irrespective of this complicated arrangement, a satisfactory result was achieved. According to Van Robbroeck (1986), this bore testimony to the goodwill and negotiating skills of the engineers on both sides. It was never necessary to take disputes between the consulting engineers to government level. Meetings at government level were only held when important policy decisions were required. Mobilisation of the study teams started in August 1983. The study was conducted in two stages (Van Robbroeck 1986).

The first stage was the identification of the layout, to be studied in detail in Stage 2. The first stage was to confirm that there were no insurmountable socio-environmental or legal barriers. It was also to establish that the benefits would be sufficient for both countries to continue with the study (Van Robbroeck 1986).

The consultants concluded the study in December 1985, and the results published in the final report in April 1986. Regarding socio-environmental impacts, the report concluded that the main impact will be on the loss of some 4000 ha of arable land and 18,700 ha of grazing, that some 1365 people will have to be resettled, and that extra employment, new and improved infrastructure, fisheries and tourism, and the distribution of the extra income from water sales in the Lesotho economy will more than offset the negative impact (Van Robbroeck 1986: 11). A cost–benefit analysis was at the order of the day regarding complex economic and socio-political issues and interests. It was this dichotomy between costs and benefits that would later become major bones of contention between interest groups and the governments, project authorities, and financiers of the project.

On institutional arrangements, the report recommended that a parastatal authority be established in each country, responsible for all the works within its own territory (LHDA and TCTA). This was due to the disproportionate size of the project to Lesotho's economy. It was therefore considered inappropriate for a binational agency to implement, maintain, and operate the project. Furthermore, and because the Vaal River water users will pay most of the cost, it was deemed necessary to establish a joint agency for monitoring and having certain powers of approval. Each country was to have equal representation on the LHWC (Van Robbroeck 1986; Vorster 1987–1988; RSA 1989; Meissner 2000e; Turton 2003a, b; Meissner 2004).

The consultants also prepared a draft treaty. This treaty contained the agreement reached at the technical level. The treaty was extensively reviewed and revised by the legal staff of the Departments of Foreign Affairs of both countries. The treaty stipulated that the benefits of the project would be divided 56–44% in Lesotho's favour. This meant that, expressed in 1986 values, by January 1995, and using 1985 prices, Lesotho would receive an estimated R1.297 billion in royalties per year (Van Robbroeck 1986) from South Africa. Nevertheless, it was not always an easy matter to organise the institutional arrangements contained in the project's treaty.

Macroconflict and Microcooperation

Events four years prior to the signing of the treaty, in October 1986, influenced the content of the treaty. The interstate conflict between the two countries reached an apex in December 1982, after South Africa launched an attack against the African National Congress (ANC) in Lesotho. In 1983, Lesotho threatened to withhold water from the project if South African military involvement continued and Lesotho maintained that it would suspend any form of cooperation concerning the project (***Die Vaderland***, 16 March 1983: 3; Sullivan 1989: 208). This was before the implementation of the project, and the situation was an indication of the macroconflict between Lesotho and South Africa.

With the military intervention in mind, Lesotho demanded that a clause be written into the treaty wherein Lesotho could shut off water in case of a political dispute. Lesotho argued that because it will deliver the water, it could also control the source. To give South Africa some reassurance, it remarked that it would inform South Africa before such a cut-off. South Africa was dissatisfied with this and demanded an uninterrupted flow in turn. South Africa also threatened Lesotho that if Lesotho should break its promise, made in a future agreement on water transfers; such a break in promise would constitute the right for military intervention (***The Daily News***, 16 April 1983: 7; Meissner 2004). Water was not the source of this dispute, but rather Lesotho's sovereign integrity, which it wanted to protect against its large neighbour. The flow of water would be a lever to influence South Africa to refrain from violating Lesotho's sovereignty.

Be that as it may, South Africa was unable to obtain such a guarantee and the negotiations over the water came to a halt (***The Star***, 12 August 1986: 11). Both South Africa and Lesotho thus used the LHWP for political gains: Lesotho to get an assurance that its territorial integrity and sovereignty would not be violated and South Africa to ensure that it would receive an uninterrupted water supply. The fact that Lesotho controlled the source of the water put the country in a powerful position to influence South Africa's behaviour.

Notwithstanding these political undertones, planning for the LHWP continued from August 1983 to August 1986 (Showers 1996). During 1984, the situation concerning the LHWP was still tense, despite (micro) technical cooperation. This was due to South Africa's unhappiness over ANC members residing in Lesotho, the presence of embassies from East bloc countries, Lesotho's critique of *apartheid*, and South Africa's assistance to the Lesotho National Liberation Army (LNLA). South Africa demanded that Lesotho enters into a security agreement with South Africa, but Lesotho declined (Barber and Barratt 1990). In response, South Africa threatened to withdraw from the LHWP if the security situation did not improve. The viewpoint held by Lesotho was that the project had nothing to do with such a treaty. South Africa, on the other hand, indicated that sabotage of the project was a possibility (***The Star***, 19 May 1984: 2; ***Rand Daily Mail***, 19 September 1984: 3; ***The Cape Times***, 22 September 1984: 2).

The incentive to be gained from the project became an important diplomatic tool for South Africa in an attempt to obtain concessions from Lesotho. Thus, the project was used to improve South Africa's external security position, indicating that South Africa still did not want to place itself in a position whereby its economic heartland would be vulnerable to decisions made by one of its 'enemies' (Leistner 1984) and thereby also jeopardising South Africa's economic security. The LHWP was, therefore, during this time seen as both a source of socio-economic development and a security concern. 'High' political concerns were not the only issues to have an impact on the situation.

On 21 September 1984, negotiations between Lesotho and South Africa took place in Cape Town. After the meeting, the project's feasibility study was restarted following the pull-out of South African engineers from the study earlier that year. The security argument was still high on South Africa's agenda. South Africa, for example, still insisted that it would not sign the treaty without an integrated security arrangement and that Lesotho should get rid of ‘political problems’, like the ANC. South Africa still felt that it could not trust Lesotho with the physical security of the project. The rationale behind this insistence was the regrettable experience it has had with the Calueque Scheme on the Kunene River in August 1975 (***Die Burger***, 22 September 1984: 5; ***Sunday Express***, 7 October 1984: 1; ***Beeld***, 9 October 1984: 14; ***Die Vaderland***, 11 October 1984: 10; Meissner 2000d).

Throughout 1984, South Africa’s security position took precedence in the issue around the LHWP. At a National Party (NP) congress, the then Prime Minister of South Africa, P.W. Botha, stated that it is difficult for South Africa to start with the project because of Lesotho’s insensitivity towards South Africa’s security needs (Coetzee 1984; Leistner 1984). The development of the Lesotho Highlands Water Project was, therefore, dependent on the international relations between the two states (Coetzee 1984). South Africa, thus, coupled the project with the issue of security throughout the 1980s. If Lesotho signed a security treaty with South Africa, it would have had a positive impact on the latter’s regional security climate. The ‘last straw’ regarding South Africa’s relations with Lesotho was the opening of the Cuban[2] embassy in Maseru and continuous support of the ANC by the Jonathan government. Late in 1985, South Africa imposed an economic blockade on Lesotho (Tsikoane 1990). This had an important although negative impact on Lesotho’s internal political situation.

Coup d’état and the Signing of the Treaty

On 16 January 1986, Gen. Maj. Lekhanya staged a *coup d’état*, toppling the Jonathan government. It was argued that South Africa was the main instigator of the *coup*, especially following evidence that South African officials had met with Lekhanya on 17 January 1986. In spite of the meeting, it could not be proved that South Africa was directly involved (***The Economist***, 25 January 1986; Baynham and Mills 1987; Sullivan 1989).

The *coup* was a watershed in the relations between South Africa and Lesotho, not only over the broad spectrum of political issues, but also regarding the water project. An ‘unfriendly’ government, according to the *apartheid* regime, was removed and replaced with a more compliant one. With the political ‘problem’ out of the way, the project could be implemented as part of South Africa’s ongoing hydraulic mission.

After the *coup*, relations between the two governments improved, especially on security and economic matters. Lesotho expatriated most of the ANC’s members and broke off diplomatic ties with communist countries, including Cuba. This improved political environment culminated in the signing of the Lesotho Highlands

[2]Due to its military support to the Angolan government against South Africa, Cuba was considered an enemy by South Africa (Barber and Barratt 1990; Meissner 2004).

Water Project Treaty on 24 October 1986 (Treaty 1986; ***Beijing Review***, 10 February 1986; Thabane 2000; Meissner 2004). The treaty was important for South Africa, assisting it to break out of its isolation mould. It was the first international treaty South Africa signed after the Nkomati Accord with Mozambique and the Lusaka Agreement with Angola in 1984 (Barratt 1985).

According to the South African government, it also showed the rest of Southern Africa that to cooperate with South Africa could have positive implications. The signing of the treaty could be seen as a reward from South Africa to Lesotho for complying with South Africa's wishes regarding the issues of the ANC and communist bloc embassies (Sullivan 1989: 209). It was therefore also a good public relations exercise for an internationally ostracised South Africa. After the signing, a series of feasibility studies were undertaken. This was to secure '…the services of engineers, natural and social scientists to investigate and report on ways in which the construction of dams and related infrastructure was going to affect people and the environment in the designated areas' (Thabane 2000: 634).

Improved Relations and Implementation: 1986–2003

From 1986 onwards, relations kept on improving. After the collapse of the Soviet Union and the unbanning of the ANC and other banned political parties in South Africa on 2 February 1990, the entire SADC region saw a meltdown of hostilities, at least in some areas. In 1992, South Africa and Lesotho exchanged diplomats, and in March 1993, the military government in Lesotho was replaced by a civilian one. Following the election of April 1993, Vincent Mokhele was sworn in as Prime Minister. He immediately committed his country to good relations with South Africa and the development of Lesotho's economy. Within this political and economic framework, the Lesotho Highlands Water Project would be of paramount importance, he remarked (***Beeld***, 31 March 1993: 15).

During the period 1993–2003, the overall international relations between South Africa and Lesotho were characterised by growing cooperation over the project. Collaboration was further strengthened by the ongoing political reforms in South Africa and the election of the ANC as the ruling party 1994. It is interesting to note that during South Africa's *apartheid* era, the ANC was against the LHWP, for political reasons. In particular, it saw the project as a 'domination instrument' by South Africa in the Southern African political arena (***The Citizen***, 16 November 1994; ***Business Day***, 23 January 1998).

The LHWP was also used by South Africa to gain leverage over the Jonathan government, to put pressure on Lesotho to get rid of the ANC and communist bloc embassies. This was unfavourably met by the ANC. After the 1994 elections, the ANC changed its position and started to support the project for the benefits the project will bring to both countries. On 22 January 1998, Phase 1A of the LHWP was put in operation. At the opening ceremony, South Africa's President Nelson Mandela and Lesotho's King Letsie III commissioned the official transfer of water to South Africa. The Namibian President and Botswana's President Sir Ketumile Masire graced the ceremony with their attendance (***Sowetan***, 22 January 1998; ***Die Volksblad***, 23 January 1998; TCTA and LHDA 2001). A year later, the Muela hydropower station was commissioned. Of Phase 1B, the Matsoku Weir and the transfer tunnels from Matsoku to Mohale and Katse dams were enabled in October 2001 (TCTA and LHDA 2001). In February 1998, the South African government decided to proceed with Phase 1B. The government claimed that it would save R500 million through this decision, which was supported by the World Bank. Bank staff concluded that only a lengthy delay could theoretically save any money on the project. Even so, this would come at an unacceptable risk of economically crippling water shortages in Gauteng by the second decade of the twenty-first century (***Business Day***, 25 March 1998: 3; ***The Star***, 25 February 1998: 3). Later that year, an event would take place that would lead to renewed interest in Lesotho's internal political affairs and the Lesotho Highlands Water Project—Operation Boleas. More on this event will follow below as it became one of many bones of contention for interest groups involved in the project.

Interest Group Involvement

Interest groups were involved in the water politics of the project before construction even started. This section of the chapter will explore their involvement. First, I will outline the issues surrounding the project they articulate (see Table 3.2).

Table 3.2 indicates that the interest groups articulated a surfeit of issues. What is also of importance is the history of their involvement and the roles interest groups played in upsetting the predictable and stable environment of decision-makers, financial institutions, and engineering contractors involved in the project.

It All Started with Faith-Based Interest Groups

A year before the signing of the Treaty, in 1985, the Mennonite Central Committee (MCC) (based in Canada and the USA) placed two field workers in the central project area for monitoring purposes. This was the start of interest group involvement in the water politics of the LHWP. Two days before the signing of the Treaty, President Semora Machel of Mozambique died after his presidential plane crashed on South African soil. Students from the National University of Lesotho reacted to his death by attempting to stage a demonstration in the capital Maseru. This demonstration was to coincide with, and to disrupt, the signing of the treaty. Police intervened and broke it up (Khits'ane 1997; Thabane 2000; Meissner 2004). After the demonstration, there was a lull in interest group activity for about 17 months, possibly due to crackdowns by the new military government.

Table 3.2 Issues and related aspects articulated by the interest groups

Issue(s) articulated	Interest group argument(s)
Water losses and higher water tariffs in the Rand Water distribution area	Interest groups are concerned that DWAF will be lax in providing education on the merits of water conservation in its haste to recover funding for the project
Resettlement of Lesotho Highland communities	According to interest groups, this is a traumatic experience for these communities
Compensation of lost land and other assets	The cash compensation package should be improved because there is an under valuation of lost gardens and trees by a factor of 10
Environmental impacts	Downstream habitats will be adversely affected by the LHWP. The fish resources and ecology of the project area have also not been mapped
Loss of resources	There is a negative impact on the quantity and quality of the natural resources in the project area, for instance pastureland decreased by 5000 ha due to the construction of Katse, Muela, and Mohale dams
Impacts due to construction	Because of construction activities, like blasting, drilling, and road construction, springs and wells dried up and villagers were forced to travel long distances to collect water
Dam safety	Drowning of humans and livestock in the Katse reservoir is a common occurrence
Impacts downstream of the project	About 150,000 people are negatively affected by reduced water flows in the Orange River
Social impacts	Social traumas include shanty towns, raising food prices (compromising food security), an increased crime rate, a higher risk of sexually transmitted diseases—HIV/AIDS included, *shebeens* (illegal bars) sprang up and prostitution became rive. All this, because of the influx of job seekers into the project area
Cultural impacts	The Katse Dam destroyed the local rain maker's ritual site, the Zionist congregation in Ha Theko lost their baptism pool, and the rain-attracting stone of Ha Tsepo was also lost due to reservoir inundation

Sources ***Business Day*** (19 March 1998: 14), Coverdale and Pottinger (1996), Archer (1996), Internet: Pottinger (1996), Horta (1996), Internet: IRN (2001a), Hoover (2001), FIVAS (2002)

In April 1988, a workshop was organised by the Transformation Resource Centre (TRC) at the request of the heads of churches in Lesotho. Part of the workshop dealt with the project. Many of the participants had close contact with the communities to be affected by the project. The workshop anticipated some of the socio-political, economic, and environmental problems likely to impact on the communities living in the project area. At the workshop, a representative committee was elected to draw up recommendations with the affected communities. These recommendations were directed by the Heads of Churches of Lesotho and presented to the Lesotho government. Consequently, a loose coalition of activists was formed to deal with the issues surrounding the communities (Khits'ane 1997; Meissner 2004). The church played an important part in the establishment of this coalition, for the church had a long-standing relationship with the Basotho and good reputation in Lesotho, dating back to the 1800s.

More Interest Groups Come to the Party

Since 1988, a number of local and international interest groups lobbied the governments, the LHDA and TCTA, and the World Bank to halt construction of the project and to implement alternatives and improved compensation. They also continued with the monitoring of the project and its impact. These interest groups hailed from different backgrounds, for example from associational (promotional) to communal interest groups. The alternatives suggested were mainly in the form of water demand management in Gauteng (Archer 1996; Meissner 2000f; Meissner 2004). Water savings in the Province would make the project obsolete, the interest groups argued. This, however, was not their only concern.

In its quarterly journal, *World Rivers Review*, the International Rivers (formerly the International Rivers Network) noted, in 1994, that the northern water transfer tunnel had to be entirely lined with concrete. This raised the cost of the project by a further R250,000,000.00 and caused a delay of one year (Coleman 1994). Publishing an article on this, the interest group attempted to articulate that the project is not as well planned as initially thought. It therefore portrayed a negative image of the project, to help sway people's opinion against it. Nonetheless, monitoring would form an important part of interest group involvement for the years to come.

Christian Aid's Tour of the Project

In January and October 1994, Robert Archer of Christian Aid first toured the project area and produced a report emphasising three concerns: the weak relationship between the Highland communities and the governments and Project authorities; the delayed implementation of the Project's compensation and development programme by the LHDA; and the difficulty the Project will face in Phase 1B if alternative good land is not found for the displaced (Archer 1996).

From 8 to 28 November 1996, Christian Aid and Oxfam, the HCAG, and the Christian Council of Lesotho again visit the Lesotho Highlands Water project. The group was assisted by other interest groups that work in the project area or have an interest therein, namely the Lesotho Council of Non-governmental Organisations, the MCC, the TRC, the Young Women's Christian Association (YMCA), and the

Group for Environmental Monitoring (GEM) (Archer 1996). The growing assistance indicates the pressing matter of the issues the communities are facing regarding the project and that the project is increasing in importance as an issue for the interest groups.

The purpose was to 'examine the LHWP's compensation programmes and other policy issues and also assessed opportunities to establish income-generating or development projects in the LHWP area' (Archer 1996). Consequently, the interest groups published wherein they outlined objectives and made recommendations. This will be discussed in more detail under the section of the chapter dealing with the interaction between the actors. Nonetheless, the visit is an indication of the seriousness with which the interest groups viewed the project's impact on the Lesotho Highlanders. This was further exemplified when authorities closed the Katse Dam's sluice gates in 1995 and the response it invoked from interest groups.

The closing sluices came only two weeks after Moea Ramokoatsi, a representative of the HCAG, met with World Bank officials in Washington, DC. The purpose of which was to request that the gates remain open until the project's critical unresolved problems were addressed (Coleman 1995). Ramokoatsi was joined by representatives from the International Rivers, Environmental Defence (ED) (based in Washington, DC), and the MCC. Ramokoatsi also met with officials of the US Treasury Department and other agencies involved in the project. The meeting's purpose was to register a list of complaints from people living near the dam. She requested that plans for Phase 1B be postponed until critical issues originating from the completed Phase 1A, affecting some 20,000 people for land lost (according to Ramokoatsi), are resolved. These issues included deteriorating health conditions and inadequate compensation (Coleman 1995).

Campaigning against the LHWP continued into 1996, when the South African Department of Water Affairs and Forestry (DWAF) reported that it might have to halt its involvement in the project after the first Phase (1A and B) was completed. This was greeted by the interest groups, in particular the International Rivers, as good news. Pottinger wrote that the LHWP is therefore 'a pipe dream of over-eager engineers'. She also stated: 'As often the case with hurriedly planned water projects in meteorologically unpredictable arid regions, the hydrological estimates were wrong: there isn't enough water to fill the planned dams, and as a result there is a lot less money for Lesotho' (Pottinger 1996; Hoover 2001: 7–9). These statements are an indication of divergent resource use perceptions between an interest group and engineers or the government benefiting from the project. The discourse of the hydraulic mission is therefore at odds with the norms articulated by individuals and interest groups.

The interest groups were not only engaging the post-apartheid South African governments, but also finding fault with the way the decision to proceed with the project was taken in the mid-1980s. Horta (1996: 20) was of the opinion that 'many project-related decisions are being made by the *apartheid*-era bureaucracy which remains entrenched in South Africa. The *apartheid*-era planners of the Lesotho Highlands Project were concerned about a reliable supply of water for industrial growth. An adequate supply of safe water for the disenfranchised black majority of the country was not a high priority'. The present South African government, although it is providing access to clean water to all, can implement alternatives to the project. These include water demand management and water savings by industries (Horta 1996). Linking the project with *apartheid* was a sure way of getting a message across. The interest groups linked the LHWP with highly controversial policy—*apartheid*, thereby linking the project to the discourse of exclusion and ostracism so in vogue under *apartheid*.

The interest groups were not merely against the LHWP, they also suggested alternatives to the project. This is a normal response when a policy is opposed. This was also the case in the aftermath of labour unrest on one of the project sites.

Response to Labour Unrest

On 14 September 1996, labour unrest broke out at the construction site of the Muela power station, near Butha Buthe. Reports state that some workers were killed by the police and that others were injured. Interest groups in Lesotho, the LCN in particular, responded strongly against the incident, condemning the police of improper conduct. The Lesotho interest groups contacted International Rivers and Environmental Defence, petitioning them to add their voices in calling for justice in the matter. Together they asked the World Bank to use its good offices to press Lesotho's government and the LHDA to take proper measures. The coalition of interest groups had even asked for an international commission of inquiry, but the government refused. It did, however, launch an internal inquiry into the matter, following pressure from the World Bank (IRN 1996a; Meissner 2000f; Meissner 2004). A World Bank team visited Lesotho in October 1996, to investigate the incident. It stated in a letter to the interest groups that, '[T]he objective of this visit was to try to determine for ourselves [World Bank team] what happened'. A report was compiled after the visit (IRN 1996a, b). This is an important indication of the transnational role and involvement of interest groups. A global development institution is lobbied by an interest group from the USA to intervene in the domestic politics of a developing African country. Not only were interest groups from the other countries actively lobbying various actors in the international political arena—local interest groups were also hard at work.

The World Bank report contains a number of recommendations. The Lesotho government should establish an independent and transparent public commission of inquiry to examine the issues that led to the labour dispute and the events. While the internal inquiry might be a useful precursor to such a commission, the establishment of the public commission of inquiry should not depend on the outcome of the internal inquiry. That the capacity of the Lesotho government be strengthened to enable it to monitor labour disputes at the project more efficiently. A dedicated unit dealing only with the LHWP might be the best option, according to the World Bank. That the LHDA should incorporate the lessons of recent events in the drafting of contracts for Phase 1B to ensure close monitoring of labour relations and occupational health and safety issues (IRN 1996a). Thus, the interest groups, through appeals to shared values (justice), convinced the World Bank to apply pressure on the government and LHDA.

Villagers File a Law Suit

Before construction commenced on the Mohale Dam in 1997, villagers from Ha Nqheku filed a lawsuit against the LHDA in the Lesotho High Court. The villagers' argument was that the LHDA was violating national laws regarding the seizure of land. The lawsuit stated that the village had had its fields, trees, and water supply negatively affected by construction work. Yet, this does not seem to be the issue. The LHDA has not registered the names of property owners in a 'book of reference' as required by Lesotho law and the project's own legal documents. The claimants asked the court to 'declare the operations of the project a violation of [their] rights' and directed the LHDA to make the books of reference available, or to stop construction if the project authorities refused. They also questioned the legality of the project's 1990 compensation regulations and asked the court to 'direct the authority to submit its accounts dealing with compensation to be inspected by [their] representatives' (Pottinger 1997b). Villagers in the Lesotho Highlands do not have access to huge amounts of financial and human resources and perceivable power. Their lawsuit is an indication to the contrary. It is also a signal that they are assisted by other interest groups, probably from a liberal democratic setting, like Canada or the USA, who transferred their knowledge of legal procedures to these communal interest groups. With the transfer of knowledge, norms and values are also communicated and by implication transferred.

Report by the Highland Church Action Group and Increased Pressure on the Governments

Two days before the commissioning of Phase 1A, in January 1998, the HCAG released a report on a survey conducted in the project area. It concluded that 75 % of the Highland villagers affected by the project felt their standard of living had decreased since the start of the project. The results also showed that 40 % of the 93 households surveyed, claimed that their grievances and compensation claims had not been addressed. Only two of the households were satisfied with compensation (***The Star***, 21 January 1998; ***Business Day***, 19 March 1998). The HCAG said: 'The inaction on the cases shows, at best, a lack of co-ordination and organisation within the [LHDA] bureaucracy. At worst, it demonstrates a lack of respect for affected people as well as a lack of co-operation with non-government organisations' (***The Star***, 21 January 1998).

Willie Croucamp, director of International Projects at the South African Department of Water Affairs and Forestry, indicated that he was satisfied that compensation was adequately addressed by the LHDA. The World Bank had set in place performance milestones for the LHDA that were successfully met. The LHDA appointed new staff to its compensation department in 1997. Croucamp furthermore remarked: 'There has been satisfactory progress for the World Bank to go ahead [partially funding Phase 1B], I think that is the best evidence that the LHDA has got behind some of the problems that have been plaguing the project. Our view is that this is not an issue (any longer)'. A survey done by the LHDA indicated that all but 14 of 679 complaints lodged in the Phase 1A area had been settled to the satisfaction of the parties (***The Star***, 21 January 1998: 5; ***Business Day***, 19 March 1998: 14).

In January 1998, after the commissioning of Phase 1A, interest groups became more vociferous in questioning the South African government's determination to continue with further phases (***Business Day***, 19 March 1998: 14). Thus, as the project's Phase 1 was completed, the lobbying intensified because of the importance of the first event and the media attention it received. On 22 January 1998, a number of interest groups released a statement, calling on the South African and Lesotho governments to halt further development of the LHWP until 'outstanding social, environmental and economic concerns' were resolved. The interest groups included GEM, ACO, and the Soweto branch of the SA National Civics Organisation, ELA, the Environmental Justice Networking Forum (EJNF), the HCAG, and International Rivers (***Business Day***, 23 January 1998: 4). The interest groups claimed that about 2000 people displaced by the Katse Dam had not been compensated. Both Lesotho and South Africa also failed to address ecological matters. These ecological concerns included the impact of reduced river flows on local fish populations, the effects of the manipulation of the natural flow canals, and the project's impact on Namibia. Water conservation would also be severely hampered by imminent increases in water supply costs. They also stated that an environmental impact assessment (EIA) had not been completed before Phase 1A commenced, which was contrary to international law and professional standards (***Business Day***, 23 January, 1998: 4; Internet: ***IPS***, 30 January 1998: 1).

The interest groups furthermore also objected to Phase 1B, particularly the construction of Mohale Dam. They argued that this phase would seriously undermine creative management efforts on the part of DWAF and Rand Water (the utility that supplies water to the metropolitan areas in Gauteng) to increase equity and efficiency in Gauteng and Southern Africa. Such efforts include education, introduction of water-saving products like dual-flush toilets, tariff reform and fixing leaks and plumbing systems, in other words water demand management. The groups moreover maintained that Phase 1B would increase water supply costs which will be passed onto the consumer (***Business Day***, 23 January, 1998: 4). The water demand management discourse was linked to strategic future considerations. Should authorities not implement water demand management in place of the LHWP, in future it will become more difficult for the roll-out of water demand management.

In a letter to the *Business Day*, Richard Sherman from GEM reiterated the 'grave concerns on ecological, social and economic grounds' of the project. He indicated that they had made their position clear about the project. From 1996 to 1997, a number of conferences and workshops were held by GEM to debate the issue. These had been attended by government officials, the World Bank, the Development Bank of Southern Africa (DBSA), and Rand Water. Sherman also wrote: 'A mountain of correspondence testifies to the well informed and urgent debate over whether the people of Lesotho have been treated properly; whether the regional ecology can stand such unprecedented reversal of water flow; whether conservation measures can now be applied (given the huge inflow of Lesotho water that must now be paid for and hence consumed); and whether the many vastly undeserved Gauteng consumers will ever receive their research [sic] [reconstruction] and development programme promise of a free, lifeline supply' (***Business Day***, 23 January 1998: 9). Thus, at the point where the public and media's interest increased over a water resources management project, interest groups will use such events to campaign more strongly against it. It is, therefore, all about propaganda and how to increase the level of attention towards the 'negative' consequences of the Lesotho Highlands Water Project.

During January 1998, the Indian interest group, Save the Narmada Movement, joined the international network of interest groups campaigning against the project. Shritad Dharmabhikary, representative of the movement, argued that they had huge successes in stopping the Sardar Sarovar Dam in the Narmada River in India. Because of this, they decided to join the network. He also had an interview with Prof. Kader Asmal, South Africa's then Minister of Water Affairs and Forestry and chair of the World Commission on Dams (WCD) (***Die Volksblad***, 28 January 1998: 2). The increased campaigning, therefore, seemed to produce results, with another knowledgeable interest group joining the coalition. This seemed to have a positive effect, especially for the Lesotho Highlanders.

Memorandum of Understanding

In May 1999, the LHDA and Lesotho interest groups signed a memorandum of understanding (MOU). This MOU, *inter alia*, addressed the responsibilities of the interest groups involved in the project. The signing thereof was a new development in the relationship between the two actor types. Since 1994, the LHDA initiated regular monthly meetings with the interest groups with whom issues of concern were discussed. Subsequently, action plans were developed and implemented. As a result, it led to the signing of the MOU, which was hailed, by both the World Bank and the UNDP as 'unique' (Meissner 2000f; Meissner 2004).

One of the most important sections of the MOU is the principle of cooperation. This principle has the task of guiding the parties' cooperative endeavours. It not only outlines the nature of the relationship between the LHDA and interest groups, but also the way interest groups will behave during their interaction with affected communities (Articles 6.1 and 6.2).

6.1 Lesotho Highlands Development Authority and the cluster of NGOs [interest groups] commit themselves to work in ways that ensure integrity, mutual respect, transparency, accountability, efficiency, full disclosure, and access to information in their dealings with each other and affected communities.
6.2 The NGOs [interest groups] commit themselves to work in ways that ensure accountability to the affected communities, integrity, effectiveness, and accountability in their implementation of specific programmes falling within the areas of cooperation identified in Sect. 5.0 of this MOU. The NGOs working on LHWP programmes that are governed by this MOU shall be capacitated to perform the services and carry out their obligations with due diligence, efficiency, and economy, in accordance with generally accepted techniques, practices, and professionalism, and shall observe sound management and technical practices (MOU 1999).

Articles 6.1 and 6.2 encompass the principles of good governance. In Section 6.5, the parties are asked to develop a code of conduct to govern the cooperative relationship, which will apply to those activities carried out on behalf of the affected communities (MOU 1999: 5). As one astute observer puts it, 'The project [LHDA] has learnt the lesson that the NGOs [interest groups] having worked with the communities have an essential role to play in the delivery of services to the communities by the project and that both the NGOs [interest groups] and the LHDA have the same objective which is to ensure delivery to the communities' (Mochebelele 2000: 111). Nevertheless, the MOU was a failure. This is reflected in the Parliamentarians' visit to the affected communities and the ombudsman inquiry. This meant that routine and institutionalised negotiations were ineffectual. A possible explanation for this is that the interest groups tried to act like states or the governmental entities that are implementing the project. The structures in which these respective organisations operate are different, not only in form but also in terms of their normative make-up. Said differently, the paradigms in which they operate are different. States function within a deeply entrenched hierarchical structure with a top-down command and control ethos. Interest groups, on the other hand, work in a more horizontal structure where the relationships between interest groups are more horizontal and a bottom-up liberal democratic ethos prevails. The MOU restricted the interest groups' freedom of movement and forced them to operate within a strict hierarchical top-down structure.

The practice of MOUs between project authorities and interest groups may become the norm in future large dam projects, and other issues affecting the environment (Meissner 2000e; Meissner 2004) and states and interest groups should at least acknowledge the fact that they operate within different structures and with dissimilar norms in hand. This is not a sure way for such MOUs to succeed, but it could waste a lot of time negotiating unrealistic MOUs. Although the MOU is an indication of routine and highly institutionalised negotiations with interest groups on the part of the LHDA, it was a failure and a more confrontational interaction ensued after its signing.

World Commission on Dams Hearings

The World Commission on Dams (WCD) was launched in February 1998. The Commission was an international and independent body consisting of 12 commissioners. The WCD dates back to April 1997, when governments, government agencies, and interest groups sponsored a meeting in Gland, Switzerland, between dam proponents and those who are critical of dams. The common ground between the two groupings was sufficient to lay the basis for the establishment of the WCD. On the establishment of the WCD, the International Rivers' Lori Pottinger said that interest groups are 'concerned about having someone [Asmal] supportive of the World Bank and a controversial dam [Katse] heading the commission'. She also stated that there should be 'now [1998] a moratorium on large dam building' as the WCD's establishment 'vindicates claims that large dams have had massively negative social, environmental and economic impacts' (***Business Day***, 20 February 1998; WCD 2001; Fujikura and Nakyama 2002). This was not to be. The impression created by International Rivers was that they were putting too much faith in an international regime.

Interaction with the World Bank

The ongoing corruption scandal that rocked the project in 1999 also led to interaction between the interest groups and the various state, parastatal, and non-state entities assisting in its implementation. The IRN and Public Services International (PSI) sent a letter to the World Bank President, James Wolfensohn, in November 1999. In this, the interest groups raised the issue of corruption and said that the Bank 'bears a responsibility here [LHWP], since it is the sponsor of large and profitable projects which attract the multinationals. The Bank has adopted clauses in its procurement guidelines which state that the Bank will declare a company ineligible for future contracts if it has engaged in corrupt practices.' They impressed on the Bank to act against the companies involved in the corruption scandal stating that the Bank 'is morally obliged to take this action. We also believe that its own guidelines, adopted to combat corruption, oblige it to do so' (IRN 1999). In another letter sent from International Rivers and Environmental Defence to the World Bank, over the issue, evoked a response from the World Bank. In the letter, they claim that the Bank plays a much larger role that mere limited funding. 'Nothing could be further from the truth. Not only did the World Bank finance the design of the project; it also is responsible for setting up and coordinating the financing program. It is unacceptable for the World Bank to claim that it is a passive bystander in the unfolding corruption investigation.' Because of this, the interest groups demanded that the Bank 'debar the companies involved in the bribery from future World Bank-financed activities. It also should launch an investigation into its own role in this controversial project' (WPC 1999). 'Arm twisting' was therefore the technique used. ACRES, one of the companies involved, were subsequently debarred from the World Bank's panel.

Responding to the letter, Jean-Louis Sarbib and Callisto Madavo from the World Bank stated that it is proud to make the project a reality, despite its limited financial contribution of 5 % of total costs. It also committed itself to fight corruption in African countries (WPC 1999). Said differently, the World Bank will not debar companies involved in the corruption scandal (disagreement was the response).

It also declared that the World supports the project because of its importance to Lesotho, South Africa, and the entire SADC region, because it sees the project serving the poor in Lesotho and South Africa (WPC 1999).

The Lesotho interest groups, in turn, responded by a letter to the World Bank's reaction stating that they, '…are troubled, however, by their [Bank officials] failure to promise World Bank sanctions against the 12 multinational corporations when it is proved that they bribed the former chief executive of the LHWP [sic.].' The interest groups also dismissed the claim that the World Bank is helping poor communities in Lesotho through a social fund set up with LHWP revenues. They called the fund, 'a tool of opportunistic politicians.' The Lesotho interest groups declared that they support the project, but question the 'openness and care with which it was prepared.' They called on the World Bank to serve the 'poor' by helping them to 'challenge the existing power and economic relations that keep [them] "poor"' (WPC 1999).

On 11 and 12 November 1999, the South African Hearings for Communities Affected by Large Dams were held in Cape Town. At these hearings, participants from Southern African countries met to discuss and analyse the negative and positive social, environmental, and economic impacts that large dams have had on their communities. The hearings were hosted by the Environmental Monitoring Group (EMG), GEM, and the Botswana Office of the International Rivers, under the patronage of Reverend Njongonkulu Ndungane, the Anglican Archbishop of Cape Town (Stott et al. 2000).

Secretariat staff of the WCD, most notably Asmal, Ronnie Kasrils, Minister of Water Affairs and Forestry, and Justice Albie Sachs of the South African Constitutional Court, were also present (Stott et al. 2000). At the hearings, villagers from the project area in Lesotho gave evidence about their experience regarding the implementation of the project and how it affected them. Didian Malisemelo Tau, from the Makotoko Village in Lesotho, told of the people's experience when Mohale Dam was constructed. He described their way of life before the LHDA asked them, in 1995, to move off their land. The village community literally lived off the land, in other words they, and according to him, lived sustainably. They had enough firewood for cooking and heating, enough clean water from springs and wells for drinking and cooking, enough pasture to raise livestock which was sold for an income, and homes. The village was therefore sustaining itself off the land, and its economy was based on subsistence farming. Maize and vegetables were grown to satisfy their own needs, and the surplus was sold (Stott et al. 2000).

The LHDA told them to move, 'because they will build a dam there which will help the R.S.A. by water. The factories and industries of R.S.A. need water to proceed their work', Tau said. The community resisted. According to Tau, 'L.H.D. A project promised to beautify our lives more than what we had. It promised to increase our buildings. They said we are few and they can help us but the people, who need our water, are many and it is difficult for them to help them because they are many. If we move away there, and the project builds a dam there, that water can save many people's lives. We agreed to move away to save many people's lives with our water and we hoped that the project will be trusted to satisfy us with all that it promised to do for us because we save many people's lives'. The community's position was, 'few people agreed to move away from a place to save many people's lives'—in other words, give a little to get a little. Similar grievances were raised by other villagers from the highlands. However, and according to the interest groups, the LHDA reneged on its policy (Stott et al. 2000).

This is an indication of what is valuable to the Highlanders and Alexandra residents (see below for more information on the Alexandra residents). There are, therefore, slight differences in the valuation of resources between the two communities, but still they loosely coalesced against a policy. The interest groups from developed countries, on the other hand, articulated issues that were more abstract and normative.

Furthermore, the interest groups criticised the LHWP in terms of non-compliance with the WCD guidelines. These guidelines are contained in the WCD's final report, published in November 2000 (IRN 2001; Fujikura and Nakyama 2001). Regarding these guidelines and the commitment made by the Lesotho and South African governments in 1986, about the resettlement and living standards of the affected communities, the IRN remarked that, 'standards of living for the majority of the project-affected people are in fact declining' (IRN 2001b). It was behind this background that the IRN criticised the LHWP against the WCD guidelines. The IRN did this by highlighting a number of issues and recommendations from the WCD with which the LHWP does not comply (see Table 3.3).

Table 3.3 World Commission on Dams guidelines and the International Rivers' criticism

WCD guideline	International rivers' criticism
Those bearing the social and environmental costs and risks of large dams are frequently not the same people who receive the social and economic benefits. The WCD, therefore, recommend that governments should give social and environmental aspects the same significance as technical, economic, and financial factors when pondering whether to build a dam	The LHWP had a profound impact on Lesotho's economy. In 1998, it accounted for 13.6 % of the GDP and royalties made up 27.8 % of government revenue. However, Lesotho's poor have seen little of these benefits. The Lesotho Highlands Water Revenue Fund (LHWRF) must distribute the royalties to the poorest in Lesotho. However, corruption forced the World Bank to restructure it
The WCD reiterates the necessity of meaningful participation of people whose livelihoods, human rights, and property and resource rights may be affected by dams. The Commission recommends negotiations in which all stakeholders have an equal opportunity to influence decisions from the beginning of the planning process	The IRN contends that participation by affected communities has been minimal at best. Affected people have had no forum to effectively negotiate how the project's dams would affect them, let alone influence the decision to build them
The WCD states that special attention is necessary to ensure that compensation and development measures are in place well in advance of resettlement. Furthermore, a clear agreement with the affected people on the sequence and stages of resettlement will be required before construction on any project preparatory work begins	Resettlement was unnecessarily stressful for LHWP-affected people. No compensation was received, despite World Bank Policy requiring it. Where resettlement took place, they were resettled to places with inadequate and unsafe drinking water and where they experienced hostility from host communities. They had no opportunity to negotiate binding performance contracts

Source IRN (2001b)

Three Individuals Complain

On 21 May 1998, the World Bank's executive board postponed, until 4 June, a decision on a US$45 million loan for Phase 1B of the project. This was after the lodging of an anonymous complaint by three residents of Alexandra (an impoverished black resident suburb of Johannesburg). In April, a similar grievance was filed by the civic organisations of Alexandra and Soweto. After a meeting with Asmal and a visit to the project, the protest was withdrawn. The three individuals decided, though, to further pursue the issue. The objection was not handled by the board, but by the inspection panel, which acts as the Bank's ombudsman. It was expected that the loan would be approved on 20 May 1998. Conversely, several members of the board wanted clarification over the implications of the complaint under the Bank's procedures. For this reason, a decision was delayed until the next available date—4 June 1998 (***Business Day***, 22 May 1998: 2; ***City Press***, 6 September 1998: 2; ***IPS***, 10 March 1998).

The complaint, supported by the IRN, stated that the Bank's analysis justifying the loan was flawed and it would make more sense to focus on WDM rather than the construction of another US$1.5 billion dam. Members of the ACO believed that the water to Gauteng would increase to such an extent that the poor would be unable to pay for it. One of the claimants commented that the Bank's delay was, 'at least an acknowledgement that [their] grievance is actually going to be addressed.' The IRN argued that the World Bank should delay its decision on the project until all social, economic, and environmental issues are resolved. Pottinger said: 'If these issues remain unresolved and the loan is approved, the World Bank will be supporting a project that is not needed for at least seven to 11 years. [This would] broadcast the message that the bank supports supply-driven water resources management, even in one of the most arid areas of the world.' On 4 June 1998, the World Bank's executive board approved the loan. Pamela Cox, World Bank country director for South Africa and Lesotho, reiterated that the project is 'the lowest cost alternative for water supply to Gauteng province' and 'a major source of development for Lesotho' (***Business Day***, 22 May 1998: 2; ***Business Day***, 5 June 1998: 3). The interest groups were, thus, unsuccessful in using the poverty issue to influence the World Bank.

After the complaint lodged by the three Alexandra individuals in May 1998, and after the World Bank's inspection panel cleared the way for no further investigations in September 1998, Asmal argued that issues raised by the claimants were 'extremely relevant.' These issues—wastage of water in townships and water tariffs—'have been vigorously pursued by the Department over the past four years.' The Minister also stated, 'It is the democratic right of individuals to question decisions of government and international organisations…[The] independent review process by the [World Bank] ombudsman proved that we are transparent in our dealings, that this is a sound project that benefits both South Africa and Lesotho…' (***City Press***, 6 September 1998: 2). The response by the Minister to 'appeals to shared values' was an alternative interpretation to the grievances of the Alexandra residents. This reaction would surface again when interest groups tried to prove that Operation Boleas was a 'water war'.

Even so, in July 1998, Jim MacNeill, a World Bank ombudsman arrived in South Africa to investigate whether the complaint by the three individuals needs further investigation. On 3 September 1998, the World Bank's inspection panel alleged that it had found no grounds for further investigation regarding the Bank violating its own policies in approving the loan. The panel agreed that water prices had risen beyond the ability of some to pay and that the leaky infrastructure was causing 'severe wastage and health problems in Alexandra, Soweto, and other townships where conditions were harsh and unsanitary for millions of people.' Yet, there was no connection between these conditions and any 'observance or not by the bank of its own policies and procedures.' The panel also found no evidence of the three individuals being intimidated by the government to drop their complaint (***Business Day***, 23 July 1998: 8; ***Business Day***, 4 September 1998: 3).

'Water Wars'

Operation Boleas was seen by the interest groups to be the first example of a water war in the Southern African region. The International Rivers, Environmental Defence, and South African Rivers Association (SARA) made a direct link between the operation and the LHWP saying, 'a massive World Bank-funded water project in the African nation of Lesotho helped spark…the type of armed confrontation water experts predict… The prejudice toward big infrastructural projects promotes unsustainable, inequitable water-management—in short, the perfect setting for future water wars' (IRN 1999c). Graeme Addison, from SARA, in a letter to the *Mail and Guardian* of 2–8 October 1998 remarked: 'The attack was more than symbolic. Like the United States in Kuwait, we [South Africa] had a strategic interest in a precious natural commodity. The Lesotho Highlands Water Project and in particular Katse Dam are the key to South African thinking (if you can call it that) about Lesotho.'

A water war is defined by Meissner (1998a, b: 20) and Turton (2003a: 112) as a violent confrontation that directly results from a desire for access to water. Water is therefore both a necessary and sufficient condition that causes a war or violent confrontation between actors.

Nonetheless, the claim by the interest groups that Operation Boleas was Southern Africa's 'first water war' was denounced by the then director general of DWAF, Mike Muller. He said that the LHWP brings benefits to both countries. In other words, a win-win situation prevails regarding the project. If South Africa did intervene in Lesotho and used the fostering of democracy as an excuse, as the conspiracy theory goes, the water from Lesotho would have become too expensive for South Africa, not only in terms of human lives, but also economically, Muller said (***Mail and Guardian***, 16 October–22 October 1998: 28). DWAF, thus, disagreed with the interest groups' 'scientific' argument that Operation Boleas was a 'water war'.

The Alexandra Civics Organisation's Continued Opposition

At a meeting of the ACO Housing Committee, held on 11 October 1998, the participants agreed to continue the campaign to oppose the Lesotho Highlands Water Project. This was after they lodged a complaint with the World Bank in May 1998 that the Bank had violated policies in pushing *apartheid* era plans to supply water to South Africa. They urged the World Bank to delay Phase 1B of the project. Their argument rested not on the disruption of their livelihoods, as in the case of the Lesotho Highlanders, but the money spent on the Mohale Dam. This, the ACO argued, would lead to water tariff increases in Gauteng and depletion of resources that could better be utilised to fix leaking pipes and taps, extend services to all residents, and create jobs (SAEP 1998). For the ACO, poverty alleviation, urban service delivery, and socio-economic development are of paramount importance. Alleviation of poverty and proper compensation are high priorities for the Lesotho Highlanders. These differences in issues arise because of the respective urban and rural location of the two interest groups. It was agreed at the meeting that the ACO would respond to the World Bank's inspection panel (which found no connection between the project and the poor state of water services in Gauteng) and build alliances with the other interest groups that are opposed to the project (Internet: SAEP 1998).

Demand for Compensation

In June 2000, people to be resettled from the project area to make way for the Matsoku Dam demanded compensation from the project authorities before they were moved. They received advice on this from their neighbours who were resettled in 1998 from the Mohale Dam site. They were advised to demand compensation before they were moved because, as the former neighbours warned: 'If you wait until you've been moved you will find you have no more power than a toothless dog' (TRC 2003).

In October 2000, the IRN reported that the Rural Development Plan, implemented in 1990 by the LHDA, had failed to electrify the homes of project-affected people. About US$1 million was set-aside for this purpose. The IRN stated: 'It appears project authorities never had any intention of following through on this commitment'. The IRN, furthermore, alleged that the project authorities admitted in December 1999 that the rural electrification programme had not yet been implemented. The IRN was sceptical whether it would ever be implemented (IRN 2000a). The IRN also criticised the project for having too many negative effects on downstream ecosystems. It is, moreover, highly inefficient for supplying water to RW's delivery area (IRN 2001b). Linking it to construct standards of behaviour, the IRN highlighted the negative aspects of the project. This is a sophisticated way of lobbying. Even so, more traditional methods of lobbying were also used.

Alternatives Suggested

A report submitted to the World Commission on Dams by the Environmental Monitoring, the Group for Environmental Monitoring and International Rivers, in November 1999, announced: 'Water conservation and demand management (WC/DM) holds tremendous potential to help the region to meet its water needs.' This was in response to show that there is an alternative to the LHWP for the alleviation of water scarcities experienced in the Rand Water delivery area (in other words Gauteng and other provinces serviced by Rand Water). The report indicated: 'Very few WC/DM measures have been implemented in southern Africa to date.

Research conducted for this report suggests that less than one-third of the 40 million urban water users who are served by developed supply systems are encouraged to use water efficiently by any measure other than escalating block tariffs' (Rothert and Macy 1999).

The report also declared: 'An increase in efficiency of only 20 % in urban and agricultural water use would save 9000 million m^3 [of water] each year—more than the combined use of Namibia, Botswana, Swaziland and Zimbabwe, and more than 10 times the combined yield of Katse and Mohale dams' (Rothert and Macy 1999). Thus, by implementing WC/DM instead of large water supply projects, like the LHWP, Southern Africa can save a large volume of water, according to the interest groups.

Demonstrators protested against the LHWP at three dam sites on 19 November 2001. In particular, they indicated the lack of fair compensation for property lost and unfulfilled promises of development in affected communities. They demanded a 10 % share of the royalties and a commission of inquiry to look into the project's impacts on local people. A petition was delivered. In it, they stated: 'We have tried by all possible means to get a fair and reasonable compensation for our property… but this was all a fiasco. We were promised development…but this has not materialised to date' (IRN 2001b). This was the second time in the project's history that a coordinated protest took place. About 1000 affected people gathered at Katse and Mohale, and about 300 marched at Muela Dam (IRN 2001b).

Mohale Reservoir and the Lakabane Family

The impoundment of the Mohale Reservoir commenced about a year later. The rising water level of the reservoir threatened the Lakabane family. A brief description of their predicament will give an indication of the nature of the interaction between a family, an interest group, and a parastatal.

The Lakabane family lived on a hill in the middle of the reservoir area. According to interest groups, when impoundment started, the family faced the prospect of drowning or relocation. The obvious option was resettlement. There was, nevertheless, a problem. The LHDA told the family that they were not entitled to resettlement because the land on which they lived and homestead was bequeathed to Makobeli, Lakabane's brother. He received letters of confirmation from the rest of the family, the district secretary and the principal chief of Thaba-Bosiu that the land and all on it belonged to Makobeli. This did not convince the LHDA (***The Survivor***, 14 November 2002).

Makobeli approached the Transformation Resource Centre to intervene and assist them in the matter. The Transformation Resource Centre made a commitment that it would negotiate with the LHDA and solve the family's problems. Other members of the community wherein the Lakabane family lived also appealed to the LHDA and the Lesotho government to resettle the family (***The Survivor***, 14 November 2002). The Transformation Resource Centre and other interest groups engaged the LHDA and the authority decided in December 2002 that the family would be resettled at Ha Tsolo near Maseru (***The Survivor***, 9 December 2002). With this, the Transformation Resource Centre intermediated between the most basic unit of society—the family—and a parastatal. This is also a good example of governance without government (e.g. Rhodes 1996) where interest groups step into assist the family.

Parliamentarians Visit the LHWP

On 7–8 November 2002, Lesotho Parliamentarians visited the areas affected by the project. The Transformation Resource Centre, through the Speaker of the National Assembly, organised the visit. Its objective was to acquaint the MPs and Senators with developments relating to the Project's social and environmental aspects (***The Survivor***, 22 January 2003).

This gave communities in the area of the Katse Dam a chance to communicate directly with their government representatives. They told them that they were disgruntled and disappointed with the way the project authorities are treating them. George Molise from the Bokong community adjacent to the Katse Dam said: 'The project has made several promises to us as far as our compensation for our communal and private assets were concerned. They used to supply us with fodder to replace our grazelands destroyed by the construction works of the project. After some time they stopped the supply of fodder saying that they would give us money. But that money has not come until now.' Another Bokong villager, Mohapi Makoetlane, testified that they do not want this dam. '[It] has brought no socio-economic developments to this area as promised' (***The Survivor***, 22 January 2003). The Bokong community appealed that a motion be put before Parliament calling for the compensation policy of the project to be made into law or at least gazette it to oblige the LHDA to comply with its provisions. The villagers indicated that the compensation policy is not legally binding (***The Survivor***, 22 January 2003) creating many loopholes. The leader of the Lesotho Workers' Party (LWP), Macaefa Billy, remarked that the plight of the communities was serious, 'heart-breaking', and needed Parliament's urgent attention. He also insisted: 'Government, through the LHDA, should be brought before the courts of law to answer all these grievances. All non-governmental organisations under the umbrella of the Lesotho Council of Non-governmental Organisations should be mobilised for funds for such a court case and solidarity' (***The Survivor***, 22 January 2003).

According the deputy leader of the Basotho Congress Party (BCP), Sekoala Toloane, it was unfortunate that MPs of the affected areas were not present to hear the grievances. He furthermore told the press: 'These are not political party issues. Nevertheless, they are national issues, which need the concerted effort of all including Parliament and government. We must lobby and fight for the establishment of a Parliamentary Portfolio Committee dealing with issues related to the affected communities and the Project. This is very important and urgent. It is the primary duty of government to ensure welfare and security of the people against projects that impact negatively on them' (***The Survivor***, 22 January 2003). These utterances are an indication that the issue of compensation had finally reached the policy agenda. However, the LHDA tried to change this.

Before the visit, the LHDA, through the Ministry of Natural Resources, informed the Transformation Resource Centre that no arrangements have been made for the Parliamentarians' visit. The LHDA also stated that such visits should be well arranged so that the MPs could be briefed by trained people and so that agreements could be made with contractors working on sites to avoid embarrassment of not allowing MPs access to some areas (***The Survivor***, 22 January 2003). In January 2003, the LHDA expressed concern that the MPs did not get a balanced brief on the project. This was after complaints by communities that they received improper compensation. The LHDA maintained that it always had an open-door policy regarding compensation matters. It also prepared a second visit to take both the Upper and Lower Houses of Parliament to the project sites where they 'will get first-hand information' (***The Survivor***, 22 January 2003). The initial visit was followed by an inquiry by the Lesotho ombudsman, so the complaints did not fall on deaf ears.

The Ombudsman Investigates

At the beginning of March 2003, the Lesotho Ombudsman, Sekara Mafisa, held a week-long formal inquiry into the complaints of the communities. The inquiry came after the ombudsman received numerous written complaints over the period from December 2002 to January 2003 from people who have been affected by the project in various ways. Seven resettled communities had the same complaints, *inter alia*, late payment of compensation money, inadequate compensation for communal assets, and no vocational training as promised. The conclusion drawn by the villagers was that the LHDA was responsible for their decreasing living standards. The ombudsman also investigated the complaints since the Lesotho government wanted to finally lay the matter of compensation to rest (***The Survivor***, 13 March 2003; ***The Survivor***, 6 August 2003; G. van der Merwe, personal communication, 20 August 2003).

Mafisa published his report at the end of July 2003. Mafisa found that the LHDA should pay interest at the commercial bank lending rate in terms of Section 39 (2) of the LHDA Order No. 23, 1986, on all compensation outstanding at the time of the inquiry. The report states: 'This delay in the payment of compensation monies subjects the already traumatized resettles/relocates to unexpected hardships' (***The Survivor***, 6 August 2003). The ombudsman furthermore recommended that the 'communities be involved truly and in earnest in the revision of the Compensation Policy' (***The Survivor***, 6 August 2003). This indicates that the communities were not completely involved in the policy process from the onset.

The report also states that compensation for communal assets should be enjoyed by both the resettled and host communities by expending the funds on development projects such as construction of access roads, water supply, electricity, and income-generating activities to avoid polarisation of communities and alienation of the resettled community by the host community. 'We endorse the idea of cooperatives as a means by which these funds may be accessed by the beneficiary communities. The communities also have *a right to suggest ideas* on how best they can access these monies without taking any risks' (emphasis added) (***The Survivor***, 6 August 2003).

Mafisa called for closer cooperation between the LHDA, the resettled communities, and the Lesotho interest groups. In the report, he stated that the LHDA is unable to function successfully in its administration of the compensation programme of the affected communities and 'do not meet it a fraction of the way.' The report, therefore, emphasised and reflected the dismal relationship between the LHDA and Lesotho interest groups. The report furthermore requested for the reparation of damaged relationships between the LHDA and interest groups. Nevertheless, according to Mafisa, the LHDA, as a public institution, should take the lead in this regard. Mafisa stated that the interest groups are good vehicles to exemplify the good image of the LHDA to the rest of the world, and this vehicle is needed by the LHDA (***The Survivor***, 6 August 2003). This is another indication that the memorandum of understanding, signed between the LHDA and interest groups, has been a failure. The ombudsman inquiry is another sign. Mafisa, therefore, recommended that the LHDA should make more use of its infrastructural power and less of its despotic power, by engaging the interest groups in a constructive manner.

The inquiry and report was a significant turning point in the relationship between the core interest groups, the affected communities, and the LHDA. The ombudsman is an officer of the state who is appointed to safeguard citizen's rights and investigate allegations of misadministration, ranging from the improper utilisation of powers to the failure to follow procedures and plain incompetence. His or her role is to enhance and not replace normal avenues of complaints like administrative courts or elected representatives. Notwithstanding the role, its investigations and findings seldom have the force of law. An ombudsman is concerned with wider administrative morality. This administrative morality concern is the gist of the change in the relationship between the three actors. In effect, the ombudsman has found that the LHDA is administrating the compensation programme in an immoral and ineffective manner. This moral aspect also surfaced when the corruption scandal rocked the project. Thus, morality does play a significant role in domestic as well as international affairs (Venter 1991; Heywood 1997).

Although the ombudsman does not have executive powers whereby the LHDA will be criminally prosecuted, the report will be used by the interest groups as a means of control over the LHDA. The interest groups view the report as a great public relations victory, which they could use as a benchmark to evaluate the performance of the LHDA and the issue of compensation. Because of this success, the interest groups might again call on the ombudsman's services. That is to say whether the interest groups and the LHDA cannot put their differences aside and recommit themselves to cooperation through a refined MOU.

Recommendations to Project Authorities

Representatives from Christian Aid, Oxfam, the Highland Church Action Group (HCAG), and the Christian Council of Lesotho met a number of South African government officials and other organisations in Pretoria, Johannesburg, and Maseru, in November 1996 (see Table 3.4 on who was met).

After the visit, a number of recommendations, contained in a report, were presented to the project authorities. The LHDA should strengthen its capacity to manage complex issues of social policy, by appointing expert staff to senior posts in the organisation and LHWC. Both governments should eliminate policy differences. The project should evaluate the impact on less-affected people of adopting a list rather than a community-based approach. It should also publicise its long-term

Table 3.4 Individuals consulted during the Christian Aid, Oxfam, the HCAG, and the Christian Council of Lesotho's visit to the LHWP in November 1996

Individuals	Organisation represented
Chief Seeiso B. Seeiso	Principal chief of Matsieng
Mr. R.T. Mochebelele	Lesotho government (LHWC)
Dr. F. Falhlbusch	Lesotho government (LHWC)
Mr. H.A. Pettenburger	South African government (LHWC)
Mr. W. Maartens	South African government (LHWC)
Mr. W. Croucamp	South African government (LHWC and DWAF)
Mr. T. Putsoane	LHDA
Mr. M. Lerotholi	Lesotho government (Ministry of Natural Resource)
Mr. T. Pekeche	Lesotho government (Ministry of Natural Resource)
Dr. M. Nyaphisi	LHDA (Environment Division)
Mr. D. Field	Hunting-Consult 4
Mr. M. Edington	Development Bank of South Africa (DBSA)
Mr. G. Wain	British High Commission
Mr. M.C. Moteane	Thaba Tseka Training Centre
Chieftaness M. Mateal	Chief of Ngoajani
Chief R. Qhobela	Chief of Muela
Chief S. Mohale	Chief of Mohale
Mr. M. Sejanamane	HCAG
Mr. Matobakele	HCAG

Source Archer (1996)

compensation policy in the project area and keep the development aspects of the compensation programme alive. The operation of the Lesotho Highlands Water Revenue Development Fund should be transparent, accountable, and politically neutral. The Lesotho government and the project authorities should use the Development Fund to spread the benefits of the project more equitably across the Highlands. The project should cease to use external contractors to build replacement houses, but local builders and local material for this purpose. The compensation programme should include all sources of income, dagga (marijuana) included. The coalition of interest groups welcomed the LHDA's policy to support work undertaken on social issues in the project area. Yet, the government, project authorities, and NGOs should publicise information about the health and social problems that are likely to occur in Phase 1B and encourage public discussion on the matter. The project should communicate its policies more clearly and consistently in the Highlands and more information in Sesotho. More NGOs could usefully work in the project area to deliver services and monitor the LHWP. The group urged those NGOs that are thinking of opening programmes to do so. The NGOs working in the project area should also form a group and seek recognition from the project. Moreover, project authorities should regularly meet NGOs that work in the area to discuss policy matters and agree guidelines for financial and other forms of cooperation between the project and NGOs (Archer 1996). The report and recommendations are part of scientific proof.

Conclusion

States played the dominant part in this materialistic and agent centric milieu. This means that they produced the international political setting within the Orange River, during the period 1956–1986. This means that they were the entities that unilaterally implemented water resources management projects and without any interference from interest groups. This is evident in the rationale behind the project, to sustain development and stimulate it. Lesotho and South Africa sought their power or utility-maximising choices or interests through, *inter alia*, the Lesotho Highlands Water Project.

This is furthermore evident in the low-level conflict between the two countries that erupted from time to time over contentious political issues, particularly *apartheid*. During such periods, the project would be used by Lesotho as a political lever. This is an important consideration, for not only was the Lesotho Highlands Water Project a potential source of water for South Africa, but it was also an instrument of influence for Lesotho.

The predominance of state actors changed when interest groups started to get involved in the water politics of the project. At first, these actors played a monitoring role, and in other words, they were watchdogs. This role changed as more components of the project were implemented. The implementation of such components led to stronger interest group opposition. With this, a new normative structure took shape around the issues of the project. One of the most significant is that large dam projects are not an absolute necessity for society to secure an abundant water supply. Alternatives, like WDM, can also be a remedy. This norm was mainly imported by outer-peripheral interest groups (ED, FIVAS, and IRN) and adopted by some of the inner-peripheral interest groups (ACO, ELA, EMG, and GEM).

The expression of this norm by the inner-peripherals was done, however, with different intentions in mind. For instance, for the ACO, development of Alexandra was the top priority. For EMG, ELA, and GEM, the environmental considerations and human rights of the Lesotho Highlanders were central.

Thus, the 'alternative-to-dams' norm is shared by interest groups, but utilised to advance different political agendas. This means that a norm is a scarce resource. By using the norm differently, interest groups play different roles, because of their dissimilar identities, ideologies, and interests. A number of interest groups clustered around this norm are most notably International Rivers, Environmental Defence, the Group for Environmental Monitoring, the Environmental Monitoring Group, Earth-Life Africa, and the Alexander Civics Organisation.

This is not the only norm around which interest groups converged. The reduction of poverty of the highlands communities and the protection of the poor is another. Here, interest groups with an ecclesiastical or philanthropic identity played a major role.

Interest groups, from both the core and outer-peripherals, attached themselves to this norm and argued against the project by being informed by the norm. Importantly, the interest groups, and their actions, are 'sustained' by these norms. These norms are the main source of information. However, these norms are also created by the interest groups, by observing the hydropolitical environment of the Lesotho Highlands Water Project. This process is called normative commensalism (a symbiotic relationship between the norms created by the interest groups and the interest groups using the norms to sustain their arguments for or against a policy, project, or programme).

References

Archer, R. 1996. *Trust in construction? The Lesotho Highlands Water Project*. London and Maseru: Christian Aid and Christian Council.

Ashton, P. 2000. *Southern Africa water conflicts: Are they inevitable or are they preventable?* Geneva: Green Cross International.

Barber, J., and J. Barratt. 1990. *South Africa's foreign policy: The search for status and security 1945–1988*. Cambridge: Cambridge University Press in association with the South African Institute of International Affairs (SAIIA).

Barratt, J. 1985. Foreign policy 1983–85: The regional context. In *South Africa: A plural society in transition*, eds. D.J. Van Vuuren et al. Durban: Butterworth Publishers.

Basson, M.S., P.H. Van Niekerk, and J.A. Van Rooyen. 1997. *Overview of water resources availability and utilization in South Africa*. Pretoria: Department of Water Affairs and Forestry.

Bath, V. 1999. Johannesburg 'corporatises' to finance city's water supply and sanitation needs. In *World of water 2000: The past, present and future*. Water and Wastewater International: Penwell Corporation.

Baynham, S. and Mills G. 1987. Lesotho: Between dependence and destabilisation. *The World Today* 43: 52–54.

Brooks, R.E.J. 1970. Oxbow and the law of international rivers. *Codicillus* 11(1): 11–13.

Brooks, D., and S. Wolfe. 2007. Water demand management as governance: Lessons from the Middle East and South Africa. In *Water resources in the Middle East: Israel-Palestinian water issues—from conflict to cooperation*, ed. H. Shuval, and H. Dweik, 311–323. Hexagon Series on Human and Environmental Security and Peace. Part VII.

Claassen, M. 2010. How much water do we have? In *A CSIR perspective on water—2010*, CSIR Report No. CSIR/NRE/PW/IR//2011/0012/A, Pretoria: Council for Scientific and Industrial Research.

Coetzee, C.F. 1984. Botswana, Lesotho and Swaziland: Relations with the RSA – (1) Lesotho (Afrikaans). *Oënskou* 2: 12–13.

Coleman, C. 1995. Shutting the gates on Lesotho: Dam moves forward, problems remain. *World Rivers Review*, 10(3). http://www.irn.org, 2 April 2003.

Conley, A.H. 1995. *A synoptic view of water resources in Southern Africa*. Paper presented at the conference of Southern Africa foundation for economic research on integrated development of regional water resources, in Nyanga, Zimbabwe.

Conley, A.H. 1996. *The need to develop the water resources of Southern Africa*. Paper presented at the Vic Falls conference on aquatic systems, July 1996, in Zimbabwe

Conley, A.H., and P. van Niekerk. 1998. Sustainable management of international waters: The Orange River case. In *The management of shared river basins: Experiences from SADC and EU*, ed. H. Savenje, and P. Van der Zaag. The Hague: Netherlands Ministry of Foreign Affairs.

Coverdale, S. and Pottinger, L. 1996. *From here to eternity: Promises versus reality on the Lesotho Highlands Water Project*. Berkeley, California: International Rivers Network. http://www.irm.org, 2 April 2003.

Department of Water Affairs (DWA). 1987. *Report of the director-general: Water affairs for the period 1 April 1987 to 31 March 1988*. Pretoria: Department of Water Affairs.

Department of Water Affairs (DWA). 2012a. *Annual Report 1 April 2011 to 31 March 2012*. Pretoria: Department of Water Affairs.

Department of Water Affairs (DWA). 2012b. *Lesotho highlands water project (LHWP)*. Pretoria: Department of Water Affairs.

Department of Water Affairs (DWA). 2013. *Reconciliation strategy for the Orange River water supply system*. Pretoria: Department of Water Affairs.

Department of Water Affairs and Forestry (DWAF). 1994. *Lesotho highlands water project*. Pretoria: Department of Water Affairs and Forestry.

Department of Water Affairs and Forestry (DWAF). 1998. *Lesotho highlands water project: Key statistics*. Pretoria: Department of Water Affairs and Forestry.

Department of Water Affairs and Forestry (DWAF). 2008. *Strategic framework on water for sustainable growth and development*. Pretoria: Department of Water Affairs and Forestry.

Department of Water Affairs and Forestry (DWAF). 2009. *Orange River: Assessment of water quality data requirements for planning purposes. Resource water quality objectives (RWQOs): Upper and lower orange water management areas (WMAs 13 and 14)*. Report No 5 (P RSA D000/00/8009/2). Pretoria: Department of Water Affairs and Forestry.

Du Pisani, A. 1992. South Africa and SADC. In *Towards a post-apartheid future: Political and economic relations in Southern Africa*, ed. G. Maasdorp, and A. Whiteside. London: Macmillan Press.

Eksteen, M.C. 1972. *Lesotho moving outwards*. (Afrikaans) Johannesburg: Perskor Publishers.

The Association for International Water and Forest Studies (FIVAS). 2002. FIVAS. http://www.solidaritetshuset.org/fivas/english/nyforsideeng.htm, 15 December 2002.

Fujikura, R. and Nakyama, M. 2003. Study on feasibility of the WCD guidelines as an operational instrument. *Water Resources Development* 18(2): 301–314.

Gleick, P.H. 1998. *The world's water 1998–1999: The biennial report on freshwater resources*. Washington, D.C.: Island Press.

Green, L.G. 1948. *To the river's end*. Cape Town: Howard B. Timmins Publishing.

Heyns, P. 2003. Water-resources management in Southern Africa. In *International waters in Southern Africa*, ed. M. Nakayama. Tokyo: United Nations University Press.

Hoover, R. 2001. *Pipe dreams: The World Bank's failed efforts to restore lives and livelihoods of dam-affected people in Lesotho*. Berkeley, California: International Rivers Network.

Hopkins, H.C. 1978. *Kakamas—from the wilderness an oasis (Afrikaans)*. Cape Town: National Book Press.

Horta, K. 1996. Making the earth rumble: The Lesotho-South African water connection. *Multinational Monitor* 17(5): 15–20.

International Rivers Network (IRN). 1996a. *Striking Lesotho dam workers killed by police: NGOs urge World Bank to take action*. Berkeley, California: International Rivers Network. http://www.irn.org, 2 April 2003.

International Rivers Network (IRN). 1996b. *LHWP labour issues—Letter to various organizations*. Berkeley, California: International Rivers Network. http://www.irn.org, 2 April 2003.

International Rivers Network (IRN). 1999. *Letter to the World Bank*, 15 November, 1999. http://www.irn.org, 2 April 2003.

International Rivers Network (IRN). 2001a. *Huge protest over large dams in Lesotho*. Berkeley, California: International Rivers Network. http://www.irn.org, 2 April 2003.

International Rivers Network (IRN). 2001b. *Lesotho Highlands Water Project, Senqu River, Lesotho*. Berkeley, California: International Rivers Network. http://www.irn.org, 2 April 2003.

Jacobs, I.M. 2012. A community in the orange: the development of a multi-level water governance framework in the Orange-Senqu river basin in Southern Africa. *International Environmental Agreements: Politics, Law and Economics* 12: 187–210.

James, L.H. 1980. Meeting the challenge of the eighties: Total water strategy needed for the Vaal Triangle. *Construction in Southern Africa*, 103–111.

Khits'ane, M. 1997. *Highlands Church Action Group's work in the areas affected by the Lesotho Highlands Water Project (LHWP)*. Berkeley, California: International Rivers Network.

Legassick, M. 1996. The will of Abraham and Elizabeth September: The struggle for land in Gordonia, 1898–1995. *Journal of African History* 37: 371–418.

Leistner, E. 1984. Lesotho: Highlands Water Project in the balance. *Africa Institute Bulletin* 24(10): 113–114.

Lesotho Highlands Development Authority (LHDA) and Trans Caledon Tunnel Authority (TCTA). 2001. *Lesotho highlands water project: Partners for life*. Information Brochure.

Lesotho Highlands Water Project (LHWP). 2003. *Lesotho highlands water project*. Maseru: Lesotho Highlands Water Project. http://www.lhwp.org , 17 May 2003.

MacDonald, W. 1913. *The conquest of the desert*. London: T.W. Laurie Ltd.

McKenzie, R.S., and C. Roth. 1994. *The evaluation of river losses from the Orange River downstream of the PK le Roux dam*. Pretoria: Water Research Commission and BKS Inc.

Meissner, R. 1998a. Piping water from the Okavango River to Namibia—the role of communities and pressure groups in water politics. *Global Dialogue*, 3(2).

Meissner, R. 1998b. *Water as 'n bron van politieke konflik en samewerking: 'n Vergelykende studie van die Midde-Ooste and Suider-Afrika*. Tesis vir die graad Magister Artium in Lettere en Wysbegeerte, Randse Afrikaanse Universiteit, Johannesburg.

Meissner, R. 2000a. The case of the Orange River. *Conflict Trends* 2: 24–27.

Meissner, R. 2000b. In the spotlight…interest groups as a role player in large water projects. *SA Waterbulletin* 26(2): 24–26.

Meissner, R. 2000c. *And now time for some good news: Cooperation in the Orange River basin*. Pretoria: African Water Issues Research Unit.

Meissner, R. 2000d. Water, a threat of war or an opportunity for peace? *Conflict Trends*, 26.

Meissner, R. 2000e. Hydropolitical hotspots in Southern Africa: The case of the Kunene River. In *Water wars: Enduring myth or impending reality*, ed. H. Solomon, and A. Turton. Durban: The African Centre for the Constructive Resolution of Disputes (ACCORD).

Meissner, R. 2000f. In the spotlight…Interest groups as a role player in large water projects. *SA Waterbulletin* 26(2): 24–26.

Meissner, R. 2004. *The transnational role and involvement of interest groups in water politics: a comparative analysis of selected Southern Africa case studies*. D.Phil Dissertation: University of Pretoria, Faculty of Humanities.

Meissner, R. 2014. Who wants to be an agent? A framework to analyse water politics and governance. *Water SA* 40(1): 1–9.

Meissner, R., and A.R. Turton. 2003. The hydrosocial contract theory and the Lesotho highlands water project. *Water Policy* 5(2): 115–126.

Memorandum of Understanding (MOU) for LHDA-NGO cooperation, 1999. Singed at Maseru, Lesotho.

Mochebelele, R.T. 2000. Good governance and the avoidance of conflicts: The Lesotho highlands water project experience. In *Water for peace in the Middle East and Southern Africa*, ed. Green Cross International. Geneva, Switzerland: Green Cross International.

Palmer, T., and A. Ainslie. 2006. *Country pasture/forage resource profiles*. Rome: Food and Agriculture Organization of the United Nations.

Pott, A., J. Hallowes, G. Backeberg, and M. Döckel. 2009. The challenges of water conservation and water demand management for irrigated agriculture in South Africa. *Water International* 34(3): 313–324.

Pottinger, L. 1996. Lesotho water project awash in troubles. *World Rivers Review*, 11(1). http://www.irn.org, 2 April 2003.

Qaddumi, H. 2008. *Practical approaches to transboundary water benefit sharing*. Working Paper 292. London: Overseas Development Institute.

Reisner, M. 1993. *Cadillac desert: The American West and its disappearing water*. New York: Penguin.

Republic of South Africa (RSA). 1962. *Report on the proposed Orange River Development Project*. Pretoria: Department of Water Affairs. Doc. No. W.P. X'62.

Republic of South Africa (RSA). 1989. *Report on the proposed revision of the levying of a charge on water supplied from the Vaal River for the benefit of the Lesotho Highlands Water Project*. Pretoria: Department of Water Affairs.

Rhodes, R.A.W. 1996. *The new governance: governing without government*, 652–667. XLIV: Political Studies.

Rosenau, J.N. 1990. *Turbulence in world politics: A theory of change and continuity*. Princeton: Princeton University Press.

Rothert, S. and Macy, P. 1999. *The potential of water conservation and demand management in Southern Africa: An untapped river*. Berkeley, California: International Rivers Network. http://www.irn.org, 2 April 2003.

Sadoff, C.W., and D. Gray. 2002. Beyond the river: the benefits of cooperation on international rivers. *Water Policy* 4: 389–403.

Schulze, R.E. 1997. *South Africa atlas of agrohydrology and—climatology*. Pretoria: Water Research Commission. Report TT82/96.

Shand, N. 1956. *Report on the regional development of the water resources of Basutoland*. Cape Town: Ninham Shand, Consulting & Chartered Civil Engineer.

Showers, K.B. 1996. Colonial and post-apartheid water projects in Southern Africa: Political agendas and environmental consequences. Paper presented at the African Studies Association meetings, San Fransisco, California, November 1996.

Smit, P. 1967a. *Lesotho: A geographic study (Afrikaans)*. Pretoria: Africa Institute.

Smit, P. 1967b. Cooperation in Southern Africa regarding power and water (Afrikaans). *Africa Institute Bulletin* 7(10): 291–299.

South Africa Yearbook, 2010/2011. 2011. *Geography and climate*. Pretoria: Government Communication and Information System.

Southern African Environmental Project (SAEP). 1998. The ACO meet to Discuss the LHWP. http://www.saep.org, 11 July, 1999.

Statistics South Africa (Stats SA). 2011. *Provinces at a glance*. Pretoria: Statistics South Africa.

Stott, N., Sack, K. and Greeff, L. 2000. *Once there was a community: Southern African hearings for communities affected by large dams*. Cape Town: Environmental Monitoring Group.

Sullivan, J. 1989. In *Southern Africa in conflict: Problems enough to share. South Africa in Southern Africa: Domestic change and international conflict* eds. E.J. Keller and L.A. Picard. Boulder, Colorado: Lynne Rienner Publishers.

Thabane, M. 2000. Shifts from old to new social and ecological environments in the Lesotho Highlands Water Scheme; Relocating residents of the Mohale Dam area. *Journal of Southern African Studies*. 26(4): 633–654.

Thompson, L., and H. Lamar. 1981. The North American and Southern African frontiers. In *The frontier in history: North America and Southern Africa compared*, ed. H. Lamar, and L. Thompson. New Haven: Yale University Press.

Trans Caledon Tunnel Authority and Lesotho Highlands Development Authority (TCTA & LHDA). 2001. *Sustainable development: Lesotho Highlands Water Project. (CD-ROM)*. Pretoria: Trans Caledon Tunnel Authority.

Trans Caledon Tunnel Authority (TCTA). 2003. *Project description, components: Phase 1A, Phase 1B*. Pretoria: Trans Caledon Tunnel Authority. http://www.metsi.com/o2/sb-02.html, 7 March 2003.

Trans Caledon Tunnel Authority (TCTA). 2013. *Lesotho highlands water project*. Pretoria: Trans Caledon Tunnel Authority. http://www.tcta.co.za/Projects/Pages/LesothoHighlands.aspx, 3 July 2013.

Treaty. 1986. *Treaty on the Lesotho highlands water project between the government of the Republic of South Africa and the government of the Kingdom of Lesotho*. Signed by representatives of both governments on 24 October 1986, Maseru Lesotho.

Tsikoane, T. 1990. *Towards a redefined role of the Lesotho Highlands Water Project in the post-apartheid Southern Africa. In Santho, S. and Sejanamane, M. (eds.), Southern Africa after apartheid: Prospects for the inner periphery in the 1990s*. Harare: SAPES.

Turton, A.R., and L. Ohlsson. 1999. *Water scarcity and social adaptive capacity: Towards an understanding of the social dynamics of managing water scarcity in developing countries*. Paper presented in the workshop no. 4: Water and social stability of the 9th Stockholm water symposium "urban stability through integrated water-related management", hosted on 9–12 August by the Stockholm Water Institute (SIWI) in Stockholm, Sweden. Also available as MEWREW Occasional Paper No. 18.

Turton, A.R. 2000. Precipitation, people, pipelines and power in Southern Africa. In *Political ecology: Science, myth and power*, ed. P. Stott, and S. Sullivan. London: Arnold.

Turton, A. 2003a. An overview of the hydropolitical dynamics of the Orange River Basin. In *International waters in Southern Africa*, ed. M. Nakyama. Tokyo: United Nations University Press.

Turton, A. 2003b. Hydropolitics: The concept and its limitations. In *Hydropolitics in the developing world: A Southern African perspective*, ed. A. Turton, and R. Henwood. Pretoria: African Water Issues Research Unit.

Turton, A.R., R. Meissner, P.M. Mampane, and O. Seremo. 2004. *A hydropolitical history of South Africa's international river basins*. Pretoria: Water Research Commission. WRC Report No. 1220/1/04.

Turton, A. 2005. *Hydro hegemony in the context of the Orange River Basin*. Paper presented at the workshop on hydro hegemony, hosted by Kings College and the School of Oriental and African studies, May 20–21, 2005, in London.

Turton, A. 2008. A South African perspective on a possible benefit-sharing approach for transboundary waters in the SADC region. *Water Alternatives* 1(2): 180–200.

Turton, A.R., and R. Meissner. 2000. *Second progress report of the institutional support task team shared river initiative on the Incomati River*. African Water Issues Research Unit (AWIRU).: Pretoria.

Van Robbroeck, T.P.C. 1986. *The Lesotho Highlands Water Project: An authoritative overview*. Pretoria: Department of Water Affairs.

Van Vuuren, L. 2012. *In the footsteps of giants—exploring the history of South Africa's large dams*. Pretoria: Water Research Commission.

Vorster, M.P. 1987–1988. The Lesotho Highlands Water Project. *South African Yearbook of International Law* 13: 95–118.

Wallis, S. 1996. *Lesotho Highlands Water Project*. London: Shani Wallis Laserline.

Warner, J. 2000. *Integrated management requires an integrated society: Towards a new hydrosocial contract for the 21*st *Century*. Pretoria, African Water Issues Research Unit (AWIRU) at the University of Pretoria. AWIRU Occasional Paper.

Warner, J.F., and R. Meissner. 2008. The Politics of Security in the Okavango River Basin: From Civil War to Saving Wetlands (1975–2002). In *International water security: domestic threats and opportunities*, ed. N.I. Pachova, M. Nakayama, and L. Jansky. Tokyo: United Nations University Press.

World Commission on Dams. 2001. *Dams and development: A new framework for decision-making. (CD-ROM)*. London: Earthscan Publications.

White, E. 1965. Last steps towards independence: The three African Protectorates. *African Affairs* 64(257): 261–270.

Wilsenach, A. 1982. Lesotho: Lesotho Highlands Water Project (Afrikaans). *Africa Institute Bulletin* 22(12): 104

The Washington Post Company (WPC). 1999. *Corruption and the World Bank: Three letters to the Washington Post on the LHWP corruption case reveal very different perspectives*. Washington D.C.: Washington Post Company. http://www.irn.org, 2 April 2003.

Chapter 4
Everyday Agency and the Lesotho Highlands Water Project

Abstract In this chapter, I put the interest groups' influence around the Lesotho Highlands Water Project into perspective. The chapter focuses on the period when interest groups started to become politically involved in the project in 1985. I furthermore discuss interest groups as agents of change. For them to operate as change agents, they need to have a certain amount of agential power, which they pit on various fronts against the agential power of states and financial institutions. I argue that agential power is dynamic, changing over time as circumstances evolve. Interest groups and governments express agential power through different norms and types of authority that characterise the political interaction between the different parties.

Keywords Agential power · Norms · Change · Influence · Subgroupism · Authority

Introduction

As I already mentioned, the Orange-Senqu River is Lesotho and South Africa's most strategic surface water resource, providing water to various users and playing an important role in the economy of the countries. For South Africa, this strategic importance was already realised in the 1960s when the South African government started implementing inter-basin transfer schemes, in other words, the transportation of water on a grand-scale to the Vaal River system to sustain the economic hub situated in South Africa's Gauteng Province. The LHWP, with its various phases, is the latest, and most probably, the last of these water transfers. This prognosis is not due only to the transnational role and involvement of interest groups. Economics and especially the availability of financial resources will in future have an impact, as well as alternative policy initiatives, articulated by interest groups, like water demand management and a move to more efficient water use practices in agriculture, the corporate environment, and our homes. Interest groups are omnipresent phenomena of any political society because of their attempts to influence public

R. Meissner, *Interest Groups, Water Politics and Governance*,
SpringerBriefs in Environmental Science, DOI 10.1007/978-3-319-21130-5_4

policy and their representation role. In other words, as long as there are policies of any type, there will be interest groups to influence these policies. These are fundamental political roles. Interest groups, because of their universal character, are involved in any policy arena, water resources management included. They are usually, but not always, involved in the domestic political arena when engaging government in the water policy sector. Over the past decade, they became transnationally involved in the Lesotho and South African water sectors through the Lesotho Highlands Water Project. Not only interest groups from South Africa are engaging governmental and parastatal institutions, but groups from around the globe are also participating. Because of their omnipresence and their persistence, their influence might not be immediately felt making water demand management as opposed to the construction of large water resources management projects a reality in future.

This chapter will capture some of these nuances, and put the transnational role and involvement of interest groups in the water politics of the LHWP into perspective. Of importance in this context is the nature of the transnational involvement of interest groups. In 1985, the MCC already had field workers in place before the Project's inception in 1986. Thus, the trantionalisation of the Project in the non-governmental domain already occurred in that year, and by an interest group with a truly transnational character. Soon thereafter, the Lesotho interest groups followed, with the establishment of the HCAG and its monitoring activities in the Project area (Meissner 2004). Interest groups from South Africa and further afield then became involved. This means that the transnational movement grew over the years to such an extent that the LHWP was no longer only an international project between two states, but a truly globalised one with the involvement of a plethora of interest groups.

This is noteworthy, for although project officials do not have a high level of routine and institutionalised negotiations with interest groups, attention was redirected away from the engineering feats to the more political and socio-economic aspects of the project. The redirection was transnational in nature, with interest groups from outside Lesotho and South Africa's borders, taking both governments and project authorities, especially the LHDA, to task about the preceived adverse impacts of the LHWP and its components. Thus, it is not only the involvement of states that lends to the LHWP a transnational character (cooperation over the flow of water across or, in this case, under the border), nor is it the involvement of contractors and subcontractors from across the world, interest groups are also responsible for this transnationalisation.

Interest Groups as Agents

Interest groups are agents within society, with a certain amount of agential power to affect policy changes (e.g. Hobson 2000). Although these agents could not stop the implementation of the Project—and the state's agential power prevailed—interest groups are part of the policy process (Meissner 2004) and by effect integrated water resources management.

The analysis of the role and involvement of interest groups in the Lesotho Highlands Water Project was also useful to indicate how the project had led to social mobilisation of the Lesotho Highlanders. The Highlanders were not used to such far-reaching changes and had to readjust their livelihoods accordingly. In the areas where the Project was implemented, the economic base changed from predominantly subsistence farming to economic activities associated with the implementation of the project. This resulted in a restructuring of social relations. For instance, not only did the affected communities increasingly come into contact with each other, but also with foreigners, workers, and interest group representatives from other countries. The ideas from the foreign interest groups can influence people's ideas of changes taking place in their lives. This can be described as the active construction of awareness among the people (Morton 2007) that were affected by the changes brought on by the project. This can be considered a sort of social construction (e.g. Weaver 2010) where the norms of those who are being interacted with change ideas and get entrenched in the normative structures of the Highlanders and ultimately influence their behaviour. The main conduit of this social construction was a critical consciousness to overcome everyday attitudes (Morton 2007) of a subsistence existence and a reaction to the changes brought on by the project.

An important component of this social construction is that the activism on the part of those lobbying against the Lesotho Highlands Water Project instilled in them a realisation that they are not victims of everyday circumstances, but active role players in the shaping of history (Morton 2007) and to a certain extent their destinies. One can only speculate over the psychological impact the project has had and will have on the affected communities. To be sure, this impact will not be uniform throughout all the communities. In this regard, more research on this issue could open a new arena in Political Science research and how psychology influences the individual in such situations. The communities reacted to the construction of the project meaning that they did not have a power base from which they could oppose the Lesotho Highlands Water Project from the onset. Their compensation can also be seen as a form of political control over individuals. This reality is sure to have some sort of psychological influence, which is beyond the scope of this book.

What is noteworthy is that the interest groups' dismal power base did not remain the same. A reconfiguration in the power relations between the Lesotho Highlanders and the other parties affected by the project took place outside the institutionalised party political system. Since independence from Britain in 1966, Lesotho was ruled for some 20 years by a single party authoritarian government and from 1986 to 1994 by a military junta. This environment is ripe for a ruling class consciousness based on domination instead of leadership that nurtures creativity and the blossoming of alternative views. The external interest groups and the faith-based associations had a fertile landscape in which they could influence communities to form another consciousness and lobby against the project authorities and governments. Put differently, the domestic and international political landscapes were conducive towards the involvement of domestic and transnational interest groups in their lobbying efforts against the project and many of its related issues and aspects (i.e., labour unrest, military intervention, perceivable sloppy planning of the project's components, and corruption).

Images and discourses played their part throughout all the lobbying campaigns. Wider domestic and international appeal was the rationale behind the decisions to mobilise interest groups both nationally and transnationally. Throughout their involvement, and despite some failures to influence the governments and project authorities effectively, the interest groups managed to tip the balance of power away from absolute domination (based on a technocratic cost-benefit paradigm) by the governments and project authorities. Whenever reacting to an issue, the interest groups initiated and effort to oppose the governments and project authorities, and to instil a sense of 'solidarity of interests' (Morton 2007) among the interest groups and between them and the affected individuals and communities.

These non-state actors' interests and inputs towards the LHWP played an insignificant role, if any role at all, when the project was first suggested in 1956. At that time, the main goal was to sustain socio-economic development in South Africa and to kick-start it in Lesotho. The only norm, within the engineering community, was that large water infrastructural projects were a necessity for development, informed by the state's hydraulic mission.

States played the dominant part in this materialistic and agent centric milieu. This means that they produced the international political setting within the Orange-Senqu River, during the period 1956 to 1986. This is evident in the rationale behind the LHWP, to sustain development and stimulate it. Lesotho and South Africa sought their power or utility-maximising choices or interests through, *inter alia*, the LHWP.

This is furthermore evident in the low-level conflict between the two countries that erupted from time-to-time over contentious political issues, particularly *apartheid*. During such periods, the Project would be used by Lesotho as a political lever. This is an important consideration, for not only was the LHWP a potential source of water for South Africa, but it was also an instrument of influence for Lesotho.

The predominance of state actors changed when interest groups started to get involved in the water politics of the project. At first, these actors played a monitoring role, in other words, they were watchdogs. This role changed as more components of the project were implemented. The implementation of such components led to stronger interest group opposition. With this, a new normative structure took shape around the issues of the project. One of the most significant is that large dam projects are not an absolute necessity for society to secure an abundant water supply. Alternatives, like water demand management, can also be a remedy. This norm was mainly imported by international interest groups (Environmental Defence, FIVAS, and International Rivers) and adopted by some of the local interest groups (Alexandra Civics Organisation, Earth Life Africa, Environmental Monitoring Group, and the Group for Environmental Monitoring).

Expression of Norms

The expression of this norm by the Lesotho and South African interest groups was done, however, with different intentions in mind. For instance, for the Alexandra Civics Organisation development of Alexandra was the top priority. For Earth Life Africa, the Environmental Monitoring Group, and the Group for Environmental Monitoring, the environmental considerations and human rights of the Lesotho Highlanders were central.

Thus, the 'alternative-to-dams' norm is shared by interest groups, but utilised to advance different political agendas. This means that a norm is a scarce resource. By using the norm differently, interest groups play different roles, because of their dissimilar identities, ideologies, and interests. A number of interest groups clustered around this norm.

This is not the only norm around which interest groups converged. The reduction of poverty of the highlands communities and the protection of the poor is another. Here, interest groups with an ecclesiastical or philanthropic identity played a major role.

Transnational and domestic interest groups attached themselves to this norm and argued against the project through this norm. Importantly, the interest groups, and their actions, are 'sustained' by the different norms. These norms are their main source of information. However, these norms are also created by the interest groups, by observing the hydropolitical environment of the LHWP. This process is called normative commensalism (a symbiotic relationship between the norms created by the interest groups and the interest groups using the norms to sustain their arguments for or against a policy, project, or programme).

Both norms are actively articulated, but only the 'protection-of-the-poor' cluster interest groups had a measure of success to inform the actions of the LHDA. The ombudsman inquiry and consequent report and World Bank reactions attest to this. The 'alternative-to-dams' norm cluster interest groups could not convince both governments and project authorities to halt the Project. The fact that there is not enough water in the Orange-Senqu River for more dams was the main consideration. Other reasons include the World Bank's low profile in financing together with South Africa's ability to finance the project from internal revenue sources. The World Bank also supported the project, which was instrumental in its implementation. Because of these reasons, the interest groups could, therefore, not lobby developed liberal-democratic governments or the World Bank to cutback funding and thereby inhibit progress of the project.

The loose coalition, the clustering of interest groups around the two main norms, and the way they engaged the governments and other non-state actors are evidence of such a community of political engagement. Regarding this, interest groups taught these actors what appropriate behaviour is, especially the LHDA, and with this brought about a number of transformations.

Because Phase 1A and 1B had been implemented and Phase 2 is under construction, it would be incorrect to say that the interest groups had failed entirely in their lobbying endeavours. Lobbying campaigns have both explicit and implicit aims. The interest groups did score some successes in changing some of the policies or the outcomes of policies, such as compensation of resettled communities and investigations into labour unrest. The interest groups failed to change or stop the Project and influence the South Africa government to embark on an absolute path of water demand management in Gauteng in favour of the Lesotho Highlands Water Project. Implicitly, the mobilisation of civil society in the Lesotho Highlands is likely to have a profound impact on the political structures within the Lesotho Highlands, if not the entire country. The changes brought about by civil society mobilisation by domestic and transnational interest groups may take on a gradual and not and immediate character.

To repeat, although the coalition of interest groups failed to stop the implementation of Phase 1A and 1B, they were responsible for a series of changes. These alterations were either broad based (within the overall South African and Lesotho water sectors) or specific (concerning the LHWP itself). What were some of the explicit and implicit results from their lobbying efforts? Their role and involvement of interest groups were unprecedented. Before the implementation of the LHWP, the Department and Water Affairs and Forestry (DWAF) (now the Department of Water and Sanitation) went about its business in an undisturbed manner. Many projects were implemented without the interference of interest groups. This is clearly not the case with the LHWP. In other words, it was no longer a matter of 'business as usual' for the Department. This change was therefore broad-based in that a governmental department had to contend with interest group involvement at an increasing rate. No longer was it only government departments, contractors, and financial institutions involved in the implementation of a project. Interest groups also started to take an interest. Civil society participation, therefore, became more pronounced. There was therefore a sea change regarding the implementation of water supply projects in the South African water sector commencing with the participation of interest groups in the LHWP. The interest groups therefore developed a more complex and rapidly changing policy environment.

This is evident in the Lesotho Ombudsman's actions. It would appear that the Ombudsman of Lesotho had his work cut out for him when he was asked to investigate the compensation policies of the LHDA. The exercise gave an opportunity for the Ombudsman office to play a role in the project and was a valuable opportunity to exercise his management abilities. The Ombudsman report resulted in the better implementation of compensation policies. There was a rethink on the part of the World Bank regarding projects of this magnitude, especially when the corruption scandal surfaced. The lobbying efforts also highlighted the plight of construction workers and their rights as employers. There was better communication between the communities and the project authorities, especially on the part of the World Bank. Along with a change in the actor dimension, the hydropolitical environment changed accordingly. No longer was the interaction between the governments of two states, financial institutions and various multi-national corporations implementing certain parts of the project. The involvement of interest groups brought about a widening of the interactive network between the state and the non-state actor communities.

Change

One of the most important and far-reaching changes was the discursive modification regarding the LHWP. Knowledge and power are inextricably linked. The one produces the other. An entity or individual extends power through the development of new types of knowledge, which is used to collect more information about something and to exercise more control over this, something. This process involves the development of discourses, and interest groups are at the forefront of this process. All-in-all, the lobbying campaigns brought about a shift from the wholesale implementation of a project based on sterile cost-benefit analyses, to an implementation of the project based on dialogue and the consideration of the human dimension. This is probably one of the most profound changes or results brought on by the interest groups: a paradigm shift from the implementation of the project based on the rational calculation of engineers and accountants to a more interpretivist and conversational rationale.

A shift in the discourse took place when interest groups started to get involved and criticised the project. No longer were water resources management projects seen as good, in providing water to a growing society. Interest groups started to question the Project's viability to society, voiced concern about its negative impacts, and even proposed an alternative policy initiative–water demand management. This meant that many citizens started to arm themselves with alternative knowledge (discourses) and became to a certain extent more powerful. Before, they took in the knowledge governments and project planners and managers provided. In the contemporary age, the citizen is more likely to question the intentions and policies of government regarding water resources management projects, and interest groups are mainly responsible for this.

Regarding this paradigm shift, Richardson (2000: 1025) states that: 'Whilst not always a threat, ideas and knowledge can have a virus-like quality and present a very real challenge to those stakeholders who have relied on the security of cocoon-like policy communities. Ideas, like viruses, tend to be destabilizing agents and demand much skill on the part of existing players, if these players are to retain their existing benefits. In practice the new ideas and their attendant policy frames often "capture" all stakeholders who then find them adjusting to a new set of rules and power distributions quite different from the old policy regimes.' Interest groups are major producers of virus-like ideas and knowledge through their agential roles. By reacting to the different developments throughout the implementation of Phase 1A and 1B, the interest groups promoted support for their demands and to put increasing pressure on the governments and project authorities. This raises an important consideration on how interest groups can express themselves to achieve a desired outcome. By branding Operation Boleas a 'water war', the LHWP is not only painted with a tarred brush, but it is also set as a 'scientific' example of some sort, that could alter thinking of social and natural scientists to conclude, rightfully or wrongfully, that it was indeed an instance where water was the direct cause of armed or violent conflict between two states. Some nine years after Operation Boleas, Likoti (2007) argued from a realist theoretical perspective that South Africa's only motivation to deploy Operation Boleas was to secure the water from the Project. Whether this interpretation of the event is correct or not, Operation Boleas will be considered by some analysts of the Project as an example of a water war. The interest groups therefore did not only had an impact on the political structures and consciousness of the Lesotho Highlands and Highlanders, but also to an extent the manner in which the epistemic community investigate the event. The interest groups, implicitly or explicitly, appealed to the identity of the epistemic community as a foundation to bring about a shift in the balance of power between the state entities and the individuals and communities.

The characteristics of change surrounding the LHWP can be summarised as follows: temporal, agential, political, and discursive. Within these domains, interest groups will (always) play a role. They are therefore, along with states and governmental institutions, the driving forces behind water politics.

Thus, interest groups are part of the policy process via their norm creating ability and their fostering of a political consciousness among those communities that were virtually powerless when the project commenced. This agential role, along with their policy shaping (influencing) and representation roles, is largely the reason why interest groups are important role players in international and domestic politics.

Furthermore, members of interest groups, such as Moea, the three Alexandra residents, and the Highland villagers, were transformed into agents by the norms they upheld and pronounced. They all got an opportunity to act. All were, therefore, a representative of the interest groups and communities they are part of. In addition, they acted politically to brought about change. Moreover, democracy upholds that they are allowed to act in the manner they did. They, therefore, used freedom of speech and association to act as agents. In other words, they embedded themselves, consciously or subconsciously, into an established and accepted international norms (freedom of speech and association) and acted from these norms.

The action and behaviour of the interest groups were made possible through inter-subjective social contexts; their actions and behaviours led to a relationship and understanding between the actors, with norms and practices responsible for this.

For instance, the interest groups' response to labour unrest, the recommendations to project authorities, the reports regarding compensation and social upliftment, the complaints by Alexandra residents, articulation of the 'water wars' discourse, interaction with the World Bank concerning corruption, and the suggestion of alternatives to the project created norms. These norms had an impact on the power relationship between the states and interest groups. The norms also gave meaning to the interest groups' action, when they were interacting with other actors regarding these issues and through the articulation of issues, like higher tariffs and the cultural impacts of the LHWP on the Highlanders.

Gramsci (1996: 60–61) notes that the immediate success of initiatives (policies, programmes, projects, lobbying campaigns, etc.) should not be the sole concern of a 'realistic politics.' This type of politics must also take on a strategic perspective by creating and safeguarding future activities. Education of people is one such activity and not only the education of the powerless but also of the powerful. It is therefore conceivable to have a situation where one has a powerless person or entity in a powerful position (Morton 2007). These powerless in seemingly powerful positions can range from government officials tasked with implementing the various policies of the project to members of the epistemic community involved in the design, implementation, and maintenance of the project (e.g. engineers, hydrologists, geologists, climatologists, physicists, botanists, geographers, and those involved in feasibility studies and environmental impact assessments). Our ability to make decisions rests on the imperfect knowledge we have of reality. This becomes clear when scientists calculated that the benefits to Lesotho outweighed the costs to the communities. This calculation was based on a rational choice platform, which in turn informed the decision to proceed with the project. Yet, as phase 1 of the project was implemented, it became apparent that the calculation was not as simple as it appeared to be at first glance. The calculation did not factor in the complexities of policy implementation at communal level and changes in the international political environment and the reaction the Lesotho Highlands Water Project would provoke from domestic and transnational interest groups. The international environment had also changed so profoundly after the collapse of the Soviet Empire that democratic principles and non-state entities would play a more prominent role in the future affairs of states than during the Cold War. In this sense, those in seemingly powerful positions were powerless in stopping interest groups becoming involved and the preceding events and changes that unfolded. Everyday politics had arrived and it was no longer the scientists that were in powerful positions to influence the historical trajectory of the LHWP. In this book, I attempted to answer Hobson and Seabrooke's (2007) organising question of 'who acts and how do their actions enable change?' The answer may seem clear; it is everyday actors, the communities in the Lesotho Highlands and the Alexandra residents, who interacted with governmental decision-makers and their structures through their bottom-up locus of agency. Following Gramsci (1996), Hobson and Seabrooke (2007), and Morton (2007), in terms of the power of the powerless and locus of agency, interest group involvement in the LHWP was not only bottom-up but also top-down, when the seemingly powerless interest groups brought about changes to the thinking, communication, and acting of state entities.

Through the articulation of such salient issues, and the roles played by the interest groups in the project, the interest groups have influenced and changed the traditional hierarchical relationship between the state and society to such an extent that it will not be the same as has been in the past. In the past, it was strictly hierarchical, with governmental decisions being implemented from the top down onto society. What has happened in the water sector in the past 20 years in Southern

Africa is that the hierarchy has in many instances fallen away or had been reversed. Where it has disintegrated is where interest groups are communicating with governmental officials and project authorities on an equal footing. Where it has turned upside down is where interest groups have been able to get their message across, and government officials or project authorities reacted to this. A clear example where this has happened was the Lesotho parliamentary visit to the project area, organised by the TRC. Thus, in a situation where interest groups are lobbying against a water supply project, there will most probably be a mixture of three 'hierarchical' situations: top-down, levelled, and bottom-up. The prevalence of one situation over the other will depend on the type of government system, and the states, and interest groups' 'agential power'. Regarding the LHWP, there were top-down, levelled and bottom-up relationships throughout the history of interest group involvement. In any event, in all three hierarchical situations, the citizen is empowered to influence governmental policies, even if it is top-down with no consultations at the onset of the policy process. As the policy is implemented, interest groups will articulate issues concerning the policy or the policy process. Thus, policy implementation is not an exclusively governmental endeavour; the citizen, through the interest group, has also staked his/her claim in the policy process.

Discussion and Conclusion

The purpose of this book is to put the political process contained in the interaction between the interest groups, governments, and project authorities around the LHWP into perspective. During the interactions between the actors during phase 1 of the project, interest groups featured more prominently in the initial lobbying activities against the project. The MCC posted two field workers in the project area in 1985. This was followed, in 1986, by the student protests in Maseru in an attempt to disrupt the signing of the Treaty. In 1988, a concerted effort was made by ecumenical interest groups to produce recommendations, together with the affected communities, regarding the impacts of the Project on these communities. Consequently, more interest groups from abroad became involved in the LHWP's water politics. Thus, it was collectivities (interest groups) that were responsible for fashioning macro-consequences. In any event, a transnational interest group (the MCC) were at first responsible for raising the awareness level of the Highland communities should the LHWP be constructed. In short, an intentional macro-input led to the transnational lobbying of interest groups against the LHWP. Thus, the opinion of collectivities was the 'tipping point' in the involvement of interest groups in the LHWP's water politics.

Moreover, interest group leaders viewed the problem differently as mass opinion in South Africa and Lesotho remained constant despite the perceived problem. In 1986, state leaders ignored the students' opinion. It was only later, when a transnational campaign against the LHWP was established that collectivity leaders (from

South Africa and the World Bank in particular) responded to the problem in terms of their own values and deliberations. This transnational action led to an increasing interaction between the interest groups, their leaders, and the governments of Lesotho and South Africa, the LHDA, the TCTA, and the World Bank.

That being the case, the interaction on the part of the interest groups took place through various control techniques, i.e. 'arm twisting', bargaining, appeals to shared values, and scientific proof (e.g. Rosenau 1990). In contrast, the state collectivities reacted to these techniques through disagreement and defiance, disputation, alternative interpretation avoidance (disinterest and apathy), and conditional agreement (e.g. Rosenau 1990). The South African government also used control techniques to further its arguments for the construction of the LHWP, i.e. appeals to shared values and scientific proof. The interest groups, on the other hand, reacted through disagreement and defiance, counterforce disputation and alternative interpretation(e.g. Rosenau 1990). Throughout the mid-1980s to 2003, these control techniques were used by both sides. These techniques are expected to be employed throughout phase 2 of the project. Even so, it is not impossible that arm twisting will be utilised less and that bargaining will come to the fore more prominently. Both the authorities and the interest groups have learned valuable lessons during phase 1, mistakes might be avoid at the behest of positive results on both sides. Even so, the findings of the Lesotho Ombudsman tipped the scale in favour of the interest groups' arguments. In other words, collectivity (state) elite assisted in the promotion of the arguments of the interest groups, because of Lesotho's changed identity, from a country with an unstable political climate to one with a stable one.

Although individuals and interest groups were responsible for starting the lobbying process against the LHWP, both these actors were of a non-state nature. Thus, individuals acting alone or in concert are the shapers of policy inputs. In this case, their initial actions were intentional leading to action be taken or avoided by government elites. In contrast, the notion of a discursive 'arms race' is not a feature of the water politics of the Lesotho Highlands Water Project, mainly because the Project's Phases 1A and 1B have been completed.

In 1985, there was one non-state actor—the MCC. During phase 1A and 1B's lobbying campaign, there were 41 state and non-state actors. This growth empowered people since the Lesotho interest groups were no longer alone in their lobbying against the project. Other interest groups were able to assist in direct communication between the leader of the HCAG, Moea Ramokoatsi, and the World Bank. Furthermore, interest groups were also able to join the loose coalition later during the campaign, most notably the various Canadian labour unions that threw in their weight behind the lobbying effort.

The interest groups are also interacting with the government of South Africa, the LHDA, the TCTA, and the World Bank regarding a number of issues concerning the LHWP. Yet this interaction has been a mixture of competition and cooperation: with the failed memorandum of understanding an indication of the cooperative endeavour between the Lesotho interests groups and the LHDA. Thus, states and their institutions are no longer the most prominent actors within the Orange River basin, because of the role and involvement of interest groups since the mid-1980s.

Since 1985, interest groups have also started to take a keen interest in the issues surrounding the water politics of the LHWP. Has there been a weakening of states in the case of the LHWP? To a certain extent yes, because they are no longer alone in the implementation process of policies in the Orange River system. Yet, the fact that phase 1A and 1B have been completed is an indication of the prominence of the state in the river system.

The reason for the mix of contestation and cooperation is probably the negative image attached to the large dam building industry, mainly because of the role and involvement of a world-wide and transnational movement against large dams. To get rid of this negative image, implementing authorities, such as the LHDA and the World Bank, will most probably cooperate with interest groups than to stand by and see how their images are tarnished by these non-state entities.

Where the South African government had been the sole actor regarding the implementation of water resources management projects on the Orange River in the past, this is no longer the case. The South African government also played an important part in the negotiation of numerous treaties regarding the governing and sharing of the Orange River with its neighbours, for instance the LHWP Treaty. Yet, its legitimacy and authority has been eroded because of the role and involvement of the plethora of interest groups in the LHWP.

Although a treaty governed the implementation of the LHWP, it was the criticism from interest groups levelled against the project that has transcended the territorial borders of both countries. The visit by Christian Aid and Oxfam in 1994 and 1996 is an indication of this process. Moea Ramokoatsi's visit to the World Bank is another example, indicating that not only the project authorities and government officials were able to speak for both countries, but private citizens as well. This process was facilitated by air transport and the personal computer. The process of 'deteritorialisation' (Rosenau 1990) is posing a challenge to the 'myth of states as sovereign actors' within the Orange River basin. The main reason for this is that interest groups are contributing to the uncertainties and contradictions over where, when, and how governments can act in a sovereign manner and under particular circumstances.

A new order has therefore emerged in the Orange River basin, with heterogeneous units interacting within its water politics. No longer is it only states, as in earlier periods, that are the main actors, interest groups have also started to raise their voices in the governing of this international river basin.

Subgroupism (Rosenau 1990) is also evident within the Orange River. The Highland communities and the Lesotho interest groups' activities are the manifestation of this process. The Highland communities have historical links with each other as well as with the interest groups because of the role the church played in some of these interest groups. Thus, they are like-minded and exist in a close-at-hand environment. Furthermore, because the interest groups were disappointed by the exclusion of the Highland communities from the decision-making process, an alienation from the governmental apparatus concerning the LHWP occurred. Due to the HCAG's establishment by the church organisations, during the 1988 workshop, there is clear evidence that subgroupism beget subgroupism. This has led to the emergence of a politics of identity among the communities and the

Lesotho interest groups. The nature of this community-based politics of identity revolves around the fact that of communities are under siege from a foreign government and a project authority, attempting to destroy the livelihoods of its members. This politics of identity has weakened the South African state and project authority's consensus around the shared goal of the LHWP. Even so, the Lesotho government reacted favourably to this politics of identity when it ordered the ombudsman enquiry. Hence, Lesotho gained some of its lost authority and legitimacy through the ombudsman process.

With respect to the nature of structures of authority, the Lesotho Highlanders resorted to coercion or the threat of the use of force to mobilise their members to campaign against the WRMPs. The reason for this is that most of the interest groups rely on voluntary membership, the exception being the Lesotho Highland communities. In sum, there is a high degree of habit of compliance within all the interest groups because they are based on voluntary association or kinship ties. Thus, the interest groups' authority is founded on informal as well as formal sources of legitimacy, coupled with a horizontal network between them and vertical hierarchies. The horizontal network is an important element in the establishment of loose coalitions. The loose coalition that formed between the interest groups is also based on the different types of authority. Here, Rosenau (2003) talks about spheres of authority that is basically different sources of informal and institutional authority that have emerged in recent decades.

Moral Authority

The various faith-based interest groups that were active in the case against the LHWP are backed by moral authority. The interest groups of note in this case are Christian Aid, the Highland Church Action Group, the MCC, and Oxfam. For these organisations, their ecclesiastical calling for the well-being of their fellow human beings are the basis of their moral authority and the habitual compliance of their members and other like-minded interest groups. The historical relationship between Lesotho and the Christian church is an important factor in this moral authority. Moshesh, the first Basotho King, had already established a close and strong relationship with French Protestant missionaries in the 1840s. This relationship with the Christian church was strengthened when Dr. John Philip, a missionary leader, visited Moshesh in 1842 (Keegan 1996). The interest groups' direct involvement in and their watchdog role are further elements in this relationship. Their particular role and involvement have also led to the establishment of a loose coalition between the ecumenical and environmental interest groups, which are focussed more on the environmental consequences of the project, thus also exhibiting an ecocentric type of moral authority. It is therefore ecumenical interest groups together with environmental interest groups that generate moral authority. This is strengthened by the salience of issues—the well-being of the affected Lesotho Highlanders.

Knowledge Authority

Interest groups themselves made use of the epistemic community and their own in-house research teams. This in-house research was used by International Rivers and other interest groups to investigate and recommend alternatives to the planned project. Christian Aid, the International Rivers Network, the MCC, and Oxfam used field workers in the affected area to research the effects the Project could have on the Highlanders. This was done in cooperation with the Lesotho interest groups, most notably the Transformation Resource Centre and the Highland Church Action Group.

Not only did the research produce limited results, but they also represented an alternative to the knowledge the governments (or more specifically engineers and other scientists) had produced to advance the projects. In both cases, a large degree of similarity therefore exists regarding the contents of the knowledge. In the case of the LHWP, it was the social integrity and the environment sustaining this social cohesion that defined the contents of knowledge.

Reputational Authority

The reputation and ability of the interest groups to influence the policy process were not lost to the World Bank. This reputation was reinforced by the past experience of the World Bank, considering that interest groups had previously been able to prevent it from assisting in the construction of other large dam projects, most notably the Sardar Sarovar Dam in India (Payne 1996). The reputation of interest groups based on their ability to emphasis the negative effects of dam building projects anywhere in the world, is the source of this type of authority. This reputational authority is also dependent on the transnational relationship between South Africa and the World Bank, making it possible for the interest groups to influence the World Bank directly.

Issue-Specific Authority

Many of the interest groups involved have become experts on the issue of water resources management projects, and the effect they have on the environment and people. This is exemplified by the fact that many of the interest groups involved, most notably International Rivers, occupied a central position in the establishment of the World Commission on Dams. Yet, the interest groups were to a certain extent ridiculed by South African government officials, for not being knowledgeable regarding the issues surrounding large dam construction, and the benefits derived from these structures. This was especially the case with foreign interest groups.

Thus, it was not so much their knowledge of a specific issue that were ridiculed, but more their interventionist type of approach to the issues at hand.

What is also noticeable is that interest groups with a certain type of issue-specific knowledge are either directly or indirectly involved. These range from interest groups with an environmental agenda, to those, such as International Rivers that campaign specifically against large dams (that are directly involved), and those with a human rights and labour agenda.

Affiliative Authority

When the churches in Lesotho convened their workshop in 1988 to discuss the likely impact of the project and the Highland communities, it was affiliative authority that induced the start of the campaign. It was this shared affiliation towards the Christian religion that led to the establishment of the Highland Church Action Group, and the rallying of other faith-based interest groups to the cause of the Lesotho interest groups. This type of authority is also sustaining the loose coalition that exists between the faith-based interest groups.

Erosion of Agential State Power

There was a substantial challenge and erosion of the national and international 'agential power' (Hobson 2000) of the states involved in the project. From a temporal dimension, previous involvement of interest groups was either limited or non-existent, except for consulting engineers and contractors. This was especially the case in the Orange River basin. In this system, there was no opposition towards the implementation of water resources management projects, before construction on the LHWP began.

The involvement and the number of interest groups in the water politics of the LHWP changed significantly during the period from the mid-1980s to the early 1990s. This was consistent with the meltdown of hostilities between South Africa, on the one hand, and other countries in Southern Africa, on the other. This benevolent aspect was conducive to the establishment of interest groups, focusing on the environment, such as the Environmental Monitoring Group, Earth Life Africa, and the Group for Environmental Monitoring, as the previous authoritarian government gave way to a more open democratic system. Hence, there was a phenomenal increase in transnational interest group activity that undermined the acceptance of the actions, and policies that are authorised at state level. This undermining was at first subnational or national, but gradually became transnational as more external interest groups started to become involved in the water politics of the water supply projects. What are of importance also are the issues that were accepted and rejected by the interest groups. It was the compensation and relocation policies of the LHDA that was the proverbial thorn in the Lesotho Highlanders' side.

Regarding the extent to which interest groups as transnational actors are influencing and bridging the traditional boundary (distinction) between the domestic and international domains, they are able to do so through the variety of roles they play. Although the domestic and international 'agential power' (Hobson 2000) of the interest groups is questionable in the Lesotho Highlands Water Project, they are *the* actors that bridged the divide between the domestic and international domains. Not even the riparian states were able to do so and in such an innovative and cost effective manner. Interest groups were able to set up a loose coalition or network within a matter of years that spanned the entire globe. This is an indication of the effectual matter with which the interest groups got into contact with like-minded others. Even so, it should be remembered that it was an interest group that started the process of the transnational involvement of the interest groups regarding the project; in 1986, the MCC had already been transnationally committed towards the project.

That may very well be, but to what extent are interest groups influencing and bridging the divide between the domestic and international domains? To answer the question, it will be necessary to determine at which system levels, interest groups operate simultaneously. The interest groups have been operating in the subnational, national, regional, and global domains concurrently. On the subnational level, the Lesotho and South African interest group interacted with each other on a constant basis. This was so with the Lesotho Highlanders and the Lesotho interest groups. Nationally, the interest groups interacted with other like-minded groups as well as with the governments. The same situation took place on the regional level. Globally, the types of actors with which the interest groups interacted were somewhat different to that on the subnational, national, and regional levels. On the global level, intergovernmental organisations, such as the World Bank, and other governments became targets of the interest groups' lobbying endeavours. Ramokoatsi's meeting with World Bank officials and the Alexandra Three's lobbying of the World Bank are indications of this global interaction. Hence, interest groups, as transnational actors, are able to influence and bridge the traditional boundary (distinction) between the domestic and international domains on the subnational, national, regional, and supranational (global) levels. Moreover, leaders from collectivities, like Ramokoatsi, at the subnational level of a country, are able, through the assistance of other interest groups, to circumvent the national and regional levels, and interact with other actors (the World Bank, European Parliamentarians, US State officials, etc.) on the global level. This holds true also for individuals who are not leaders, like the Alexandra Three. An important aspect in this process is the involvement of interest groups, usually but not always from developed countries, to financially assist those interest groups from developing countries to circumvent the national and regional levels. This process is called circumventionism: the deliberate circumvention of the national and regional levels to directly lobby actors on the global level (see Fig. 4.1).

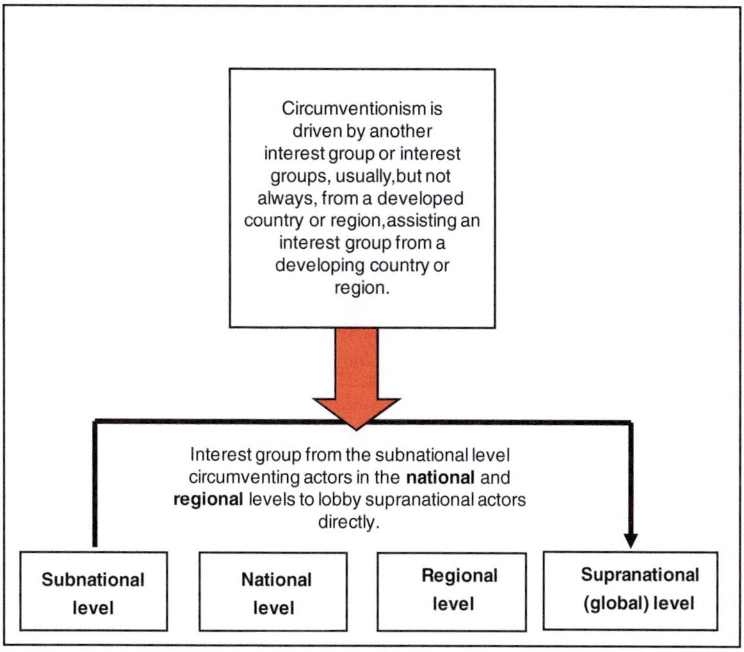

Fig. 4.1 Circumventionism

Circumventionism is an important element in the relationship between interest groups and governments regarding the implementation of policies or programmes. Interest groups are willing to sacrifice a state's sovereignty, through circumventionism, by lobbying other governments or international organisations such as the World Bank, for these actors to get involved in an internal matter on the interest group's behalf. In this way, there is not a distinction of the boundary between the domestic and international domains any longer; it has completely disintegrated. This means that the territorial integrity of the state has also been compromised. Interest groups are active and unfeeling when they sacrifice a state's sovereignty and territorial integrity when bridging the boundary, and influencing states and other types of actors outside a country's borders. What this also means is that interest groups will do everything in their power to articulate an issue, even if such an action costs the state its sovereignty and territorial integrity.

Following this, norms are also drivers of the interactive process between interest groups and governments or other state and non-state actors. This interaction is linked to the process of normative commensalism. Normative commensalism assists interest groups to learn what is happening in the hydropolitical environment and to influence governments, international organisations, and other non-state entities. Because influence is a component of power relationships, norms and norm creation act like facilitators within the interactive process. Thus, when considering the processes of circumventionism and normative commensalism, interest groups and norms are active in the realm of world politics to bring about different interactive approaches. Put in another way, interest groups, as actors, and norms, as abstract shared values, are independent variables in the political process of international river basins.

In any event, with the disappearance of the divide between the domestic and international political domains, there has also occurred a change, by interest groups, in the traditional hierarchical relationship between state and society (government and the citizen). Interest groups have mainly done this through the erosion of habitual compliance towards policies. To be more precise, the different roles interest groups' play is the causes of this changed hierarchical relationship between state and society. Through their three generic roles: discursive, participation, and philanthropic, interest groups have come to fill an emptiness left by governments. This void is constantly recreated, as governments are unable to fulfil their duties towards their citizens, such as the provision of wholesome freshwater and a healthy and safe environment to live in. From the case study, it is clear that the governments of the states and the project authorities have reneged on their duty to 'look out' for the needs of the affected communities and the natural environment. This was not done deliberately, though. Even so, interest groups readily stepped into take up the plight of the Lesotho Highlanders and to lobby for the betterment of their living conditions.

References

Gramsci, A. 1996. *The prison notebooks*, Vol. 2. Ed. J.A. Buttigieg (trans: J.A. Buttigieg). New York: Columbia University Press.

Hobson, J.M. 2000. *The state and international relations*. Cambridge: Cambridge University Press.

Hobson, J.M., and L. Seabrooke. 2007. Everyday IPE: revealing everyday forms of change in the world economy. In *Everyday politics of the world economy*, ed. J.M. Hobson, and L. Seabrooke. Cambridge: Cambridge University Press.

Keegan, T. 1996. *Colonial South Africa and the origins of the racial order*. London: Leicester University Press.

Likoti, F.J. 2007. The 1998 military intervention in Lesotho: SADC peace mission or resource war? *International Peacekeeping* 14(2): 251–263.

Morton, A.D. 2007. Global capitalism and the peasantry in Mexico: The recomposition of class struggle. *The Journal of Peasant Studies* 34(3–4): 441–473.

Meissner, R. 2004. *The transnational role and involvement of interest groups in water politics: a comparative analysis of selected Southern Africa case studies*. D.Phil Dissertation: University of Pretoria, Faculty of Humanities.

Payne, R.A. 1996. Non-profit environmental organizations in world politics: Domestic structure and transnational relations. *Policy Studies Review*, 14 (Spring/Summer 1995): 171–182.

Richardson, J. 2000. Government, interest groups and policy change. *Political Studies* 48: 1006–1025.

Rosenau, J.N. 1990. *Turbulence inhap world politics: A Theory of change and continuity*. Princeton: Princeton University Press.

Rosenau, J.N. 2003. *Distant proximities: Dynamics beyond globalization*. Princeton: Princeton University Press.

Weaver, C. 2010. The strategic social construction of the World Bank's gender and development policy norm. In *Owing development: Creating policy norms in the IMF and the World Bank*, ed. S. Park, and A. Vetterlein. Cambridge: Cambridge University Press.

MIX
Papier aus verantwortungsvollen Quellen
Paper from responsible sources
FSC® C105338

If you have any concerns about our products,
you can contact us on
ProductSafety@springernature.com

In case Publisher is established outside the EU,
the EU authorized representative is:
Springer Nature Customer Service Center GmbH
Europaplatz 3, 69115 Heidelberg, Germany

Printed by Libri Plureos GmbH
in Hamburg, Germany